THE INTERPRETATION OF QUANTUM MECHANICS

THE UNIVERSITY OF WESTERN ONTARIO
SERIES IN PHILOSOPHY OF SCIENCE

A SERIES OF BOOKS

ON PHILOSOPHY OF SCIENCE, METHODOLOGY,

AND EPISTEMOLOGY

PUBLISHED IN CONNECTION WITH

THE UNIVERSITY OF WESTERN ONTARIO

PHILOSOPHY OF SCIENCE PROGRAMME

VOLUME 3

JEFFREY BUB

University of Western Ontario, Ontario, Canada, and
Institute for the History and Philosophy of Science, Tel Aviv University, Israel

THE INTERPRETATION
OF QUANTUM MECHANICS

D. REIDEL PUBLISHING COMPANY

DORDRECHT-HOLLAND / BOSTON- U.S.A.

Library of Congress Catalog Card Number 74–76479

Cloth edition: ISBN 90 277 0465 1
Paperback edition: ISBN 90 277 0466 X

Published by D. Reidel Publishing Company,
P.O. Box 17, Dordrecht, Holland

Sold and distributed in the U.S.A., Canada, and Mexico
by D. Reidel Publishing Company, Inc.
306 Dartmouth Street, Boston,
Mass. 02116, U.S.A.

Printed in The Netherlands by D. Reidel, Dordrecht

TABLE OF CONTENTS

PREFACE   VII

I.	The Statistical Algorithm of Quantum Mechanics	1
	I. Remarks	1
	II. Early Formulations	3
	III. Hilbert Space	8
	IV. The Statistical Algorithm	15
	V. Generalization of the Statistical Algorithm	24
	VI. Compatibility	28
II.	The Problem of Completeness	32
	I. The Classical Theory of Probability and Quantum Mechanics	32
	II. Uncertainty and Complementarity	36
	III. Hidden Variables	46
III.	Von Neumann's Completeness Proof	49
IV.	Lattice Theory: The Jauch and Piron Proof	55
V.	The Imbedding Theorem of Kochen and Specker	65
VI.	The Bell-Wigner Locality Argument	72
VII.	Resolution of the Completeness Problem	84
VIII.	The Logic of Events	92
	I. Remarks	92
	II. Classical Logic	93
	III. Mechanics	105
IX.	Imbeddability and Validity	108
X.	The Statistics of Non-Boolean Event Structures	119
XI.	The Measurement Problem	128
XII.	The Interpretation of Quantum Mechanics	142

BIBLIOGRAPHY   151

INDEX OF SUBJECTS   153

PREFACE

This book is a contribution to a problem in foundational studies, the problem of the interpretation of quantum mechanics, in the sense of the theoretical significance of the transition from classical to quantum mechanics.

The obvious difference between classical and quantum mechanics is that quantum mechanics is statistical and classical mechanics isn't. Moreover, the statistical character of the quantum theory appears to be irreducible: unlike classical statistical mechanics, the probabilities are not generated by measures on a probability space, i.e. by distributions over atomic events or classical states. But how can a theory of mechanics be statistical and complete?

Answers to this question which originate with the Copenhagen interpretation of Bohr and Heisenberg appeal to the limited possibilities of measurement at the microlevel. To put it crudely: Those little electrons, protons, mesons, etc., are so tiny, and our fingers so clumsy, that whenever we poke an elementary particle to see which way it will jump, we disturb the system radically – so radically, in fact, that a considerable amount of information derived from previous measurements is no longer applicable to the system. We might replace our fingers by finer probes, but the finest possible probes are the elementary particles themselves, and it is argued that the difficulty really arises for these. Heisenberg's γ-ray microscope, a thought experiment for measuring the position and momentum of an electron by a scattered photon, is designed to show a reciprocal relationship between information inferrable from the experiment concerning the position of the electron and information concerning the momentum of the electron. Because of this necessary information loss on measurement, it is suggested that we need a new kind of mechanics for the microlevel, a mechanics dealing with the dispositions for microsystems to be disturbed in certain ways in situations defined by macroscopic measuring instruments. A God's-eye view is rejected as an operationally meaningless abstraction.

Now, it is not at all clear that the statistical relations of quantum mechanics characterize a theory of this sort. After all, the genesis of quantum mechanics had nothing whatsoever to do with a measurement problem at the microlevel, but rather with purely theoretical problems concerning the inadequacy of classical mechanics for the account of radiation phenomena. Bohm and others have proposed that the quantum theory is incomplete, in the sense that the statistical states of the theory represent probability distributions over 'hidden' variables. Historically, then, the controversy concerning the completeness of quantum mechanics has taken this form: A majority view for completeness, understood in the sense of the disturbance theory of measurement, and a minority view for incompleteness.

An interpretation of quantum mechanics should show in what fundamental respects the theory is related to preceding theories. I propose that quantum mechanics is to be understood as a 'principle' theory, in Einstein's sense of the term. The distinction here is between principle theories, which introduce abstract structural constraints that events are held to satisfy (e.g. classical thermodynamics), and constructive theories, which aim to reduce a wide class of diverse systems to component systems of a particular kind (e.g. the molecular hypothesis of the kinetic theory of gases). For Einstein, the special and general theories of relativity are principle theories of space-time structure.

I see quantum mechanics as a principle theory of logical structure: the type of structural constraint introduced concerns the way in which the properties of a mechanical system can hang together. The propositional structure of a system is represented by the algebra of idempotent magnitudes – characteristic functions on the phase space of the system in the case of classical mechanics, projection operators on the Hilbert space of the system in the case of quantum mechanics. Thus, the propositional structure of a classical mechanical system is isomorphic to the Boolean algebra of subsets of the phase space of the system, while the logical structure of a quantum mechanical system is represented by the partial Boolean algebra of subspaces of a Hilbert space. In general, this is a non-Boolean algebra that is not imbeddable in a Boolean algebra. As principle theories, classical mechanics and quantum mechanics specify different kinds of constraints on the possible events open to a physical system, i.e. they define different possibility structures of events.

This view arises naturally from the Kochen and Specker theory of partial Boolean algebras, which resolves the completeness problem by properly characterizing the category of algebraic structures underlying the statistical relations of the theory. Kochen and Specker show that it is not in general possible to represent the statistical states of a quantum mechanical system as measures on a classical probability space, in such a way that the algebraic structure of the magnitudes of the system is preserved. Of course, the statistical states of a quantum mechanical system can be represented by measures on a classical probability space if the algebraic structure of the magnitudes is *not* preserved. But such a representation has no theoretical interest *in itself* in this context. The variety of hidden variable theories which have been proposed all involve some such representation, and are interesting only insofar as they introduce new ideas relevant to current theoretical problems. Invariably, the reasons proposed for considering a new algebraic structure of a specific kind are plausibility arguments derived from some metaphysical view of the universe, or arguments which confuse the construction of a hidden variable theory of this sort with a solution to the completeness problem.

I reject the Copenhagen disturbance theory of measurement and the hidden variable approach, because they misconstrue the foundational problem of interpretation by introducing extraneous considerations which are completely unmotivated theoretically, and because they stem from an inadequate theory of logical structure. With the solution of the completeness problem, all problems in the way of a realist interpretation of quantum mechanics disappear, and the measurement problem is exposed as a pseudo-problem.

The short bibliography lists only works directly cited, and since the sources of the ideas discussed will be obvious throughout, I have not thought it necessary to introduce explicit references in the text, except in the case of quotations.

THE STATISTICAL ALGORITHM OF
QUANTUM MECHANICS

I. REMARKS

Classical mechanics – Newton's general theory of motion developed and articulated by Euler, Lagrange, and Hamilton – describes the temporal evolution of a mechanical system in terms of the change in certain appropriate physical magnitudes (e.g. energy, angular momentum, etc.), which are represented as real-valued functions on a 'phase space' X, a linear (vector) space parametrized by generalized position and momentum coordinates, the phase variables.

For a free particle, the phase space is 6-dimensional, with position coordinates q_1, q_2, q_3, representing the location of the particle in space, and corresponding momentum coordinates p_1, p_2, p_3. The classical mechanical equations of motion, Hamilton's equations:

$$\frac{dq_i}{dt} = \frac{\partial H}{\partial p_i} \qquad \frac{dp_i}{dt} = -\frac{\partial H}{\partial q_i} \qquad (i = 1, 2, 3)$$

determine a trajectory in phase space, given the initial values of the variables q_1, q_2, q_3; p_1, p_2, p_3. The quantity H, the Hamiltonian, is a function of the phase variables and characterizes the particular system involved. Since the physical magnitudes are functions of the phase variables, the values of these quantities are defined for every point on the phase trajectory of the system, and are determined for all time via Hamilton's equations by any point on the trajectory, i.e. by an assignment of values to the 'canonically conjugate' sets of phase variables q_1, q_2, q_3 and p_1, p_2, p_3. Such an assignment of values – the specification of a point $x = (q_1, q_2, q_3; p_1, p_2, p_3)$ in phase space – is a classical mechanical *state*.

Electromagnetic phenomena were incorporated into this scheme by the Faraday-Maxwell theory of fields, a field being something like a mechanical system with a continuous infinity of phase variables. This extension of classical mechanics began to collapse towards the end of the 19th century. The quantum theory was conceived in 1900 with Max Planck's solution to the 'ultra-violet catastrophe': Planck proposed that electromagnetic radiation is emitted and absorbed in discrete 'quanta', each energy quantum being proportional to the frequency of the radiation. The birth of quantum mechanics followed after a gestation period of 25 years, in the dual form of Schrödinger's wave mechanics and the Heisenberg-Born-Jordan matrix mechanics. Schrödinger demonstrated the equivalence of the two theories, and a unified 'transformation theory' was developed by Dirac and Jordan on the basis of Born's probabilistic interpretation of the wave function.

In Section II, I sketch the basic ideas behind matrix mechanics and wave mechanics, presenting these theories as different algorithms for generating the set of possible energy values of a system. I discuss von Neumann's critique of the Dirac-Jordan transformation theory, and show that matrix and wave mechanics are equivalent in the sense that they represent formulations of a mechanical theory in terms of different realizations of Hilbert space. The exposition in this section follows von Neumann. My purpose is to show the origin of the Hilbert space formulation of quantum mechanics.

I develop the geometry of Hilbert space in Section III. I introduce the notion of a Hilbert space as a vector space over the field of complex numbers, with a scalar product which defines the metric in the space.

The core of this chapter is Section IV. Quantum mechanics incorporates an algorithm for assigning probabilities to ranges of values of the physical magnitudes. I introduce this algorithm in the elementary form applicable to the finite-dimensional case. Essentially, probabilities are generated by statistical states according to a certain rule. The 'pure' statistical states are represented by the unit vectors in Hilbert space; the physical magnitudes are represented by operators associated with orthogonal sets of unit vectors, corresponding to the possible 'quantized' values of the magnitudes. These orthogonal sets of unit vectors function like Cartesian coordinate systems in a Euclidean space. The probability assigned by a particular vector, ψ, to the value a_i of the magnitude A is given by the square of the projection of ψ onto the unit vector α_i ('Cartesian axis') corresponding to the value a_i. The problem of 'degenerate' magnitudes – magnitudes associated with m possible values and m orthogonal vectors in an n-dimensional space $(m < n)$ – involves a generalization of the statistical algorithm, which is dealt with by the apparatus of projection operators and subspaces. The infinite-dimensional case requires a further generalization, in terms of the 'spectral measure' of an operator representing a physical magnitude.

Finally, the possibility of statistical states representing 'mixtures' of pure states involves a generalization in terms of the notion of the 'trace' of an operator in Section V. This version of the statistical algorithm represents the probability assigned by the statistical state W to the range S of the magnitude A as the trace of the product $WP_A(S)$, where $P_A(S)$ is defined by the spectral measure of the operator representing A for the Borel set S.

Chapter I concludes with some remarks on the compatibility relation defined on the set of magnitudes, corresponding to the commutativity of the corresponding Hilbert space operators.

II. EARLY FORMULATIONS

Both matrix and wave mechanics propose algorithms for generating the set of possible energy values of a system, and the transition probabilities between the corresponding 'stationary states'. For simplicity, consider a 1-dimensional example, say a particle confined to one dimension of space.

The method of matrix mechanics characterizes a quantum mechanical system corresponding to a classical mechanical system with the Hamiltonian function $H(q, p)$ by a Hamiltonian matrix $H(Q, P)$, i.e. the classical phase variables q, p are associated with certain matrices Q, P, and the classical Hamiltonian is associated with a corresponding Hamiltonian matrix. A matrix is simply an element of a certain non-commutative algebra, which can be represented by an array of (complex) numbers with a finite or countable number of rows and columns. Different representations are possible for the same matrix. The matrices Q, P are required to satisfy the commutation relation

$$QP - PQ = ih/2\pi$$

where h is Planck's constant and $i = \sqrt{-1}$.

Under certain conditions, which we assume satisfied, there is a representation in which the numbers in the array representing H are all zero, except along the diagonal, where they take real values. These are the possible energy values of the system. (The position of a number in an array representing a matrix is identified by the row number and column number. By a diagonal element of a square matrix, I mean an element in the position: row-i, column-i, for any i. The off-diagonal elements are those in positions: row-i, column-j, $i \neq j$.)

The arrays representing the matrices Q, P in this representation determine the transition probabilities according to a certain rule. This representation is found by solving the 'eigenvalue equation' for H:

$$\sum_j H_{ij}\varepsilon_j = e\varepsilon_i$$

where H_{ij} represents the element (complex number) in row-i and column-j of the array in some arbitrary representation. The eigenvalue equation will generally have a countable number of distinct solutions, i.e. there will be a countable number of distinct 'eigenvalues'

$$e_1, e_2, \ldots$$

and associated 'eigenvectors'

$$(\varepsilon_1^{(1)}, \varepsilon_2^{(1)}, \ldots), (\varepsilon_1^{(2)}, \varepsilon_2^{(2)}, \ldots), \ldots$$

which satisfy the equation, under the condition $\sum_i |\varepsilon_i^{(k)}|^2 < \infty$, for each k (where $|\varepsilon_i^{(k)}|$ denotes the absolute value of the complex number $\varepsilon_i^{(k)}$). Here each eigenvector $(\varepsilon_1^{(k)}, \varepsilon_2^{(k)}, \ldots)$ is a sequence of numbers, representing the components of the kth eigenvector in the representation. The superscript (k) refers to the corresponding eigenvalue.

The diagonal representation of H is obtained as the product

$$S^{-1}HS$$

where S is the matrix whose columns are the eigenvector solutions to the eigenvalue equation for H. The elements along the diagonal in this representation are the eigenvalues, i.e. the eigenvalues of H are the possible energy values of the system. Thus, the algorithm of matrix mechanics reduces the problem of generating the set of possible energy values of a system to the eigenvalue problem.

To sum up: The possible energy values of a system are obtained as the diagonal values of the Hamiltonian matrix H of the system in a certain privileged representation. H, expressed as an array of numbers with respect to any given initial representation, is diagonalized – transformed to this privileged representation – by a matrix S, constructed from the eigenvector solutions to the eigenvalue equation in the initial represtion. The matrices Q and P, transformed to the representation in which H is diagonal by the transformation matrix S, determine the transition probabilities between the stationary states corresponding the the different possible energy levels.

In the wave mechanical formulation, a quantum mechanical system corresponding to a classical mechanical system with the Hamiltonian function $H(q, p)$ is characterized by a differential functional operator $H(q, -ih/2\pi(\partial/\partial q))$ in 'configuration space', i.e. the space parametrized by the position coordinates of the system. The possible energy values of the system are those values of e for which the differential equation

$$H\left(q, - ih/2\pi \frac{\partial}{\partial q}\right) \psi(q) = e\psi(q)$$

has solutions satisfying the condition $\int_{-\infty}^{\infty} |\psi(q)|^2 \, dq < \infty$. This is Schrödinger's wave equation, an eigenvalue equation for the eigenvalues e and corresponding (complex-valued) eigenfunctions ψ. The eigenvalues e are the possible energy values, and the eigenfunctions represent the corresponding stationary states.

The Dirac-Jordan transformation theory exploits an apparent analogy between the 'continuous' configuration space parametrized by the variable q in the case of the wave functions $\psi(q)$, and the 'discrete' space parametrized by the index i in the case of the sequences $(\varepsilon_1, \varepsilon_2, ...)$, regarded as functions $\varepsilon(i)$ of the variable i. That is, a particular sequence $(\varepsilon_1, \varepsilon_2, ...)$ is regarded as a map ε from a space of integers $(1, 2, ...)$ into the complex numbers, and it is proposed that the equivalence of matrix mechanics and wave mechanics has to do with some structural similarity between this space and the space R of real numbers which the wave functions map into the complex numbers. On this view, summation over the 'discrete' variable i of the countable space corresponds to integration over the 'continuous' variable q of the uncountable space, and so the eigenvalue equation of matrix mechanics

$$\sum_{i'} H_{ii'}\varepsilon_{i'} = e\varepsilon_i$$

should correspond to an eigenvalue equation of the form

$$\int H(q, q')\, \psi(q')\, \mathrm{d}q' = e\psi(q)$$

in the case of wave mechanics. But the eigenvalue equation of wave mechanics is

$$H\left(q, -ih/2\pi\, \frac{\partial}{\partial q}\right)\psi(q) = e\psi(q)$$

and, as von Neumann points out in the introduction to his classic treatise *Mathematical Foundations of Quantum Mechanics*, it is not possible in general to represent a differential functional operator as an integral operator, i.e. in general a function $H(q, q')$ such that

$$\int H(q, q')\, \psi(q')\, \mathrm{d}q' = H\left(q, -ih/2\pi\, \frac{\partial}{\partial q}\right)\psi(q)$$

does not exist.

Such a function $H(q, q')$ is referred to as the 'kernel' of the differential operator H. The Dirac-Jordan transformation theory in effect assumes the existence of kernels for the differential operators of wave mechanics, when in fact no such functions exist. The kernel of the identity operator is represented by the Dirac δ-function, $\delta(q-q')$, so that

$$\psi(q) = \int_{-\infty}^{\infty} \delta(q - q')\, \psi(q')\, \mathrm{d}q',$$

and the kernels of other differential operators are expressed in terms of the δ-function and its derivatives. But the δ-function is an impossible map from the real numbers into the complex numbers: it would have to map every point except q onto zero, and still satisfy

$$\int_{-\infty}^{\infty} \delta(q - q')\, \mathrm{d}q' = 1.$$

In fact, the real 'analogy' is not between the countable space of points i and the uncountable space of points q, but between the space of sequences $(\varepsilon_1, \varepsilon_2, \ldots)$ with finite metric $\sum_i |\varepsilon_i|^2$ and the space of functions $\psi(q)$

with finite metric $\int_{-\infty}^{\infty} |\psi(q)|^2 \, dq$. These two spaces are isomorphic, in the sense that there exists a one-one mapping h from the function space onto the sequence space which preserves linearity

$$h(a\psi + b\varphi) = ah(\psi) + bh(\varphi)$$

and the metric, i.e. if $h(\psi) = (\varepsilon_1, \varepsilon_2, \ldots)$, then:

$$\int_{-\infty}^{\infty} |\psi(q)|^2 \, dq = \sum_i |\varepsilon_i|^2 .$$

(More generally, the scalar product is preserved. If $h(\psi) = (\varepsilon_1, \varepsilon_2, \ldots)$, $h(\varphi) = (\eta_1, \eta_2, \ldots)$, then

$$\int_{-\infty}^{\infty} \psi(q)^* \, \varphi(q) \, dq = \sum_i \varepsilon_i^* \eta_i .$$

The * denotes the complex conjugate.)

Matrix mechanics and wave mechanics are therefore equivalent in the following sense: The formulation of matrix mechanics involves a linear space of sequences with finite metric; wave mechanics is formulated in terms of a linear space of functions with finite metric. These two spaces are isomorphic. Now, the matrices Q and P – operators in the sequence space – satisfy the same commutation relations as the functional operators q and $-ih/2\pi(\partial/\partial q)$, and these commutation relations are preserved under the isomorphism h. It follows that the matrices Q and P correspond to the functional operators q and $-ih/2\pi(\partial/\partial q)$, respectively, under the isomorphism. Since the Hamiltonian matrix $H(Q, P)$ and the Hamiltonian functional operator $H(q, -ih/2\pi(\partial/\partial q))$ are constructed by the same algebraic operations from the matrices Q, P and the functional operators $q, -ih/2\pi(\partial/\partial q)$, and these operations are preserved under the isomorphism, $H(Q, P)$ and $H(q, -ih/2\pi(\partial/\partial q))$ correspond to one another under the isomorphism. Thus, matrix mechanics and wave mechanics are mathematically equivalent theories formulated in terms of isomorphic representations.

The algebraic structure of which the sequence space of matrix mechanics and the function space of wave mechanics are particular realizations is *Hilbert space*. Von Neumann develops the quantum mechanical description of events in terms of this structure. Thus, von Neumann's contribution is not merely a rigorous reformulation of the Dirac theory that avoids the mathematical difficulties of the δ-function, but a formulation of quantum mechanics that abstracts what is common to matrix mechanics and wave mechanics in an essential way. The Dirac theory misrepresents the sense in which matrix mechanics and wave mechanics are equivalent, and so the structure underlying these theories is not properly identified.

In the following sections of this chapter, I shall present an exposition of the elements of quantum mechanics, following von Neumann. I shall not be concerned with the details of particular realizations, which are relevant only insofar as they facilitate calculations in concrete problems.

III. HILBERT SPACE

A Hilbert space $\mathcal{H}$ is a linear space over the field of complex numbers. This means that if $\alpha, \beta \in \mathcal{H}$ then $\alpha + \beta \in \mathcal{H}$, and $c\alpha \in \mathcal{H}$ where c is any complex number. The elements of a linear space are called vectors. Vector addition is associative and commutative. Multiplication by complex numbers is distributive (i.e. $c(\alpha + \beta) = c\alpha + c\beta$; $(c_1 + c_2)\alpha = = c_1\alpha + c_2\alpha$) and associative (i.e. $c_1(c_2\alpha) = (c_1c_2)\alpha$). Notational convention: Unless otherwise specified, lower case Greek letters label vectors, i.e. elements of the space; lower case Latin letters label (complex) numbers.

A *scalar product* is defined on $\mathcal{H}$: a map from the set of ordered pairs of vectors, $\mathcal{H} \times \mathcal{H}$, into the complex numbers, satisfying certain conditions. The image (complex number) of the pair of vectors α, β is denoted by (α, β). The conditions on the map are:

$(\alpha, \beta) = (\beta, \alpha)^*$
$(\alpha, \beta + \gamma) = (\alpha, \beta) + (\alpha, \gamma)$
$(\alpha, c\beta) = c(\alpha, \beta)$
$(\alpha, \alpha) > 0$ unless $\alpha = 0$, the null vector, in which case $(\alpha, \alpha) = 0$

The *norm* (i.e. 'magnitude') of a vector is defined as $\|\psi\| = \sqrt{(\psi, \psi)}$. The real-valued function on $\mathcal{H} \times \mathcal{H}$ that associates each pair of vectors ψ, φ with the norm of their difference $\|\psi - \varphi\|$ is a metric on the Hilbert space $\mathscr{H}$. It follows from the properties of the scalar product that

$$\|\psi - \varphi\| \geqslant 0$$
$$\|\psi - \varphi\| = 0 \quad \text{if and only if} \quad \psi = \varphi$$
$$\|\psi - \varphi\| = \|\varphi - \psi\|$$
$$\|\psi - \varphi\| \leqslant \|\psi - \chi\| + \|\chi - \varphi\|.$$

(The 'triangle inequality' is evidently equivalent to the inequality $\|\psi + \varphi\| \leqslant \|\psi\| + \|\varphi\|$, and this follows from the Schwartz inequality $|(\psi, \varphi)| \leqslant \|\psi\| \|\varphi\|$, where $|(\psi, \varphi)|$ denotes the absolute value of the complex number (ψ, φ).)

Thus, the length or magnitude of a vector ψ is $\|\psi\|$, and the distance between points ψ and φ in $\mathscr{H}$ is $\|\psi - \varphi\|$. It is consistent with this definition of the distance function to take the condition $(\psi, \varphi) = 0$ as the condition of orthogonality of ψ and φ. (E.g., Pythagoras' theorem holds for the length of the hypotenuse of a right-angled triangle.)

The *dimension* of the space $\mathscr{H}$ is defined via the concept of *linear independence*: A set of vectors $\alpha, ..., \alpha_n$ is linearly independent if there is no way for a linear combination

$$c_1 \alpha_1 + c_2 \alpha_2 + ... + c_n \alpha_n$$

to sum to the null vector, other than by all c_i being equal to zero, i.e. if $c_1 \alpha_1 + c_2 \alpha_2 + ... + c_n \alpha_n = 0$ only if $c_i = 0$ for all i, where the c_i are any complex numbers. If n vectors are mutually orthogonal, i.e. $(\alpha_i, \alpha_j) = 0$, $i \neq j$, then they are linearly independent. For suppose

$$\sum_{i=1}^{n} c_i \alpha_i = 0$$

then

$$\left(\alpha_j, \sum_{i=1}^{n} c_i \alpha_i \right) = (\alpha_j, 0) = 0 \quad (j = 1, ..., n).$$

But

$$\left(\alpha_j, \sum_{i=1}^{n} c_i \alpha_i \right) = c_j (\alpha_j, \alpha_j)$$

and so

$$c_j (\alpha_j, \alpha_j) = 0$$

i.e.

$$c_j = 0 \quad (j = 1, ..., n).$$

A *linear manifold* in $\mathscr{H}$ is a set of vectors such that $c\alpha$ and $\alpha + \beta$ belong to the set if α, β belong to the set and c is any complex number. Equivalently, a linear manifold contains all linear combinations of any finite subset of its elements. The linear manifold spanned by a set of vectors is the smallest linear manifold that includes the set, i.e. it contains all and only linear combinations of vectors from the set. The dimension of a linear manifold is the maximum number of linearly independent vectors in the manifold. If the linear manifold contains n linearly independent vectors, but every set of $n+1$ vectors is linearly dependent, then it is n-dimensional. If there is no maximum number, i.e. if there are arbitrarily many linearly independent vectors, then the manifold is infinite-dimensional.

An n-dimensional Hilbert space $\mathscr{H}_n$ is an n-dimensional Euclidean space over the field of complex numbers. Since an orthogonal set $\{\alpha_1, ..., \alpha_n\}$ is linearly independent, it follows that the set $\{\psi, \alpha_1, ..., \alpha_n\}$, for any vector $\psi \in \mathscr{H}_n$, is linearly dependent, i.e. that there exist complex numbers $k_1, k_2, ..., k_n$, not all zero, such that

$$k\psi + k_1\alpha_1 + ... + k_n\alpha_n = 0$$

or

$$\psi = c_1\alpha_1 + ... + c_n\alpha_n \quad (c_i = -k_i/k).$$

Thus, any vector ψ can be expressed in terms of its components relative to a 'Cartesian coordinate system' or set of axes, i.e. as a linear combination of its projections along any set $\{\alpha_i\}$ of n mutually orthogonal unit vectors in $\mathscr{H}_n$.

The condition of orthogonality requires that

$$(\alpha_i, \alpha_j) = 0, \quad i \neq j.$$

In addition, each α_i is of unit length, i.e.

$$\|\alpha_i\|^2 = (\alpha_i, \alpha_i) = 1.$$

It follows that

$$(\alpha_i, \psi) = c_i(\alpha_i, \alpha_i) = c_i$$

and so

$$\psi = (\alpha_1, \psi)\alpha_1 + (\alpha_2, \psi)\alpha_2 + ... + (\alpha_n, \psi)\alpha_n$$

i.e. any vector ψ in $\mathscr{H}_n$ can be expressed in the form

$$\psi = \sum_{i=1}^{n} (\alpha_i, \psi) \, \alpha_i .$$

The set $\{\alpha_i\}$ is said to be a complete orthonormal set in $\mathscr{H}_n$, an orthogonal set of unit vectors spanning the whole space.

Notice that the absolute value of the scalar product, $|(\alpha_i, \psi)|$, is the length of the projection of ψ along α_i, and that

$$\|\psi\| = \sqrt{(\psi, \psi)} = \sqrt{\sum_{i=1}^{n} |(\alpha_i, \psi)|^2} :$$

the square of the length of ψ is the sum of the squares of the lengths of the components of ψ along n orthogonal axes. This is Pythagoras' theorem in n-dimensional Hilbert space. The difference between $\mathscr{H}_n$ and a real n-dimensional Euclidean space is that the components of a vector in $\mathscr{H}_n$ are in general complex numbers, so that, for example, $|(\varphi, \psi)|$ rather than (φ, ψ) – a complex number – is the magnitude of the projection of ψ along φ.

In $\mathscr{H}_\infty$ problems of continuity and convergence become important, but these aspects of the infinite-dimensional case are not really relevant to the problem of interpretation. I include the following discussion of terminology for completeness.

A sequence of vectors $\{\psi_i\}$ in $\mathscr{H}_\infty$ converges to ψ, and ψ is the limit of the sequence, if the sequence of numbers $\{\|\psi - \psi_i\|\}$ converges to zero. A function $f: \mathscr{H}_\infty \to \mathscr{H}_\infty$ is continuous at the point φ if for each $e > 0$ there exists a $d > 0$ such that $\|f(\psi) - f(\varphi)\| < e$ if $|\psi - \varphi| < d$. Of course, these definitions apply also in the case of $\mathscr{H}_n$. Generally, it is assumed that $\mathscr{H}_\infty$ satisfies the conditions of *separability* and *completeness* (not to be confused with completeness in the sense of the problem of hidden variables). Separability is the property that there exists a countable set in $\mathscr{H}_\infty$ which is everywhere dense in $\mathscr{H}_\infty$, i.e. every vector in $\mathscr{H}_\infty$ is the limit of a convergent sequence from the countable set or, equivalently, for any $\psi \in \mathscr{H}_\infty$ there is an element ψ_n in the countable set such that $\|\psi - \psi_n\| < e$, for any positive real number e. Completeness is the property that every convergent sequence in $\mathscr{H}_\infty$ converges to a limit point in $\mathscr{H}_\infty$ (i.e. every sequence satisfying the Cauchy convergence criterion that for each $e > 0$ there exists an $N(e)$ such that $\|\psi_m - \psi_n\| < e$ if $m, n > N(e)$).

A finite-dimensional Hilbert space is necessarily separable and complete.

A *subspace* $\mathscr{K}$ of a Hilbert space $\mathscr{H}$ is a closed linear manifold, i.e. a linear manifold containing all its limit points. Evidently, $\mathscr{K}$ is itself a Hilbert space, with dimension less than or equal to the dimension of $\mathscr{H}$. The set of vectors in $\mathscr{H}$ orthogonal to all vectors in $\mathscr{K}$ is a subspace $\mathscr{K}^{\perp} = \mathscr{H} - \mathscr{K}$. Two subspaces $\mathscr{K}_1$ and $\mathscr{K}_2$ are orthogonal if every vector in $\mathscr{K}_1$ is orthogonal to every vector in $\mathscr{K}_2$, i.e. if $\mathscr{K}_1 \subseteq \mathscr{K}_2^{\perp}$ or, equivalently, if $\mathscr{K}_2 \subseteq \mathscr{K}_1^{\perp}$.

It is a theorem that a linear manifold generated by a sequence of vectors is also spanned by a countable orthonormal set of vectors. (Von Neumann, Theorem 8.) In fact, there is a constructive procedure for generating the orthonormal set from the original set of vectors, the Schmidt orthonormalization procedure. It is also a theorem that every subspace is spanned by a countable orthonormal set. In particular, $\mathscr{H}$ is spanned by a countable orthonormal set. (Von Neumann, Theorem 9.) The property of separability is crucial here. Thus, the major consequence of the assumption of separability is the existence of Cartesian coordinate systems in the sense of *countable* sets of basis vectors in terms of which any vector can be represented. If $\{\alpha_i\}$ is a complete orthonormal set in $\mathscr{H}_{\infty}$, then

$$\psi = \sum_{i=1}^{\infty} (\alpha_i, \psi)\, \alpha_i$$

in the sense that the sequence

$$\sum_{i=1}^{n} (\alpha_i, \psi)\, \alpha_i$$

converges to ψ. Of course, in $\mathscr{H}_n$ a complete orthonormal set of vectors is always finite.

If $\mathscr{K}$ is a subspace of $\mathscr{H}$, then each vector ψ can be resolved in one and only one way into two components

$$\psi = \psi_1 + \psi_2 = P\psi + P^{\perp}\psi$$

where $\psi_1 = P\psi \in \mathscr{K}$ is the projection of ψ in the subspace $\mathscr{K}$, and $\psi_2 = P^{\perp}\psi \in \mathscr{K}^{\perp}$ is the projection of ψ in the subspace $\mathscr{K}^{\perp}$. The map $P \colon \mathscr{H} \to \mathscr{H}$ associating each vector in $\mathscr{H}$ with its projection in $\mathscr{K}$, is a special case of a *linear operator* in $\mathcal{H}$, i.e. the conditions

$$P(\psi + \varphi) = P\psi + P\varphi$$

and

$$P(c\psi) = c(P\psi)$$

are satisfied. It is obvious geometrically that P is *idempotent*, i.e. $P^2 = P$ in the sense that $P(P\psi) = P\psi$, and *self-adjoint*, i.e.

$$(P\psi, \varphi) = (\psi, P\varphi) \quad \text{for all} \quad \psi, \varphi \in \mathcal{H}.$$

(If

$$\psi = P\psi + P^\perp\psi = \psi_1 + \psi_2$$

and

$$\varphi = P\varphi + P^\perp\varphi = \varphi_1 + \varphi_2$$

then

$$(P\psi, \varphi) = (\psi_1, \varphi) = (\psi_1, \varphi_1)$$

and

$$(\psi, P\varphi) = (\psi, \varphi_1) = (\psi_1, \varphi_1).)$$

A *projection operator* may be defined as a self-adjoint, idempotent linear operator on $\mathcal{H}$, and it is then a theorem that each projection operator corresponds to a unique subspace which is its range. (Von Neumann, Theorem 12.) It is an obvious consequence of the definition that

$$(\psi, P\psi) = (\psi, P^2\psi) = (P\psi, P\psi) = \|P\psi\|^2$$

i.e. the square of the length of the projection of ψ onto a subspace is $(\psi, P\psi)$, where P is the projection operator corresponding to the subspace.

In general, an operator A is a function from a subset of $\mathcal{H}$ into $\mathcal{H}$. The operator A is said to be *defined everywhere* if the domain of A is $\mathcal{H}$. Projection operators are defined everywhere. Notice that the operator $A \cdot A = A^2$ is defined only if the range of A is contained in the domain of A. This is the case for projection operators. An operator A is linear if its domain is a linear manifold and if

$$A(a_1\alpha_1 + a_2\alpha_2 + \ldots + a_n\alpha_n) = a_1 A\alpha_1 + a_2 A\alpha_2 + \ldots + a_n A\alpha_n$$

for any linear combination of n vectors in the manifold. It follows that the range of A is also a linear manifold.

For completeness, again, I include the following brief account of linear operators in Hilbert space (until the end of this section), although these details are not of importance for the concerns of this inquiry.

A linear operator is *bounded* if, for all $\psi \in \mathcal{H}$, $\|A\psi\| \leqslant c\|\psi\|$, where $0 \leqslant c < \infty$. It is *continuous* at the point ψ if $A\psi_n \to A\psi$ whenever $\psi_n \to \psi$. If A is bounded it is continuous. Conversely, if A is continuous at one point (say the point $\psi = 0$), it is continuous everywhere and bounded. Hence the concepts of continuity and boundedness coincide for linear operators. (Von Neumann, Theorem 18.) In $\mathcal{H}_n$ every linear operator is bounded.

A linear operator A' is an *extension* of the linear operator A if the domain of A is included in the domain of A', and $A'\psi = A\psi$ for all ψ in the domain of A. If $\{\psi_n\}$ is a Cauchy sequence in the domain of a bounded linear operator A and $\psi_n \to \psi$, where ψ is not in the domain of A, then A has a unique extension to the operator A' defined on the closure of the domain of A (i.e. the smallest subspace that includes the domain of A) by the condition $A\psi = \lim_{n\to\infty} A\psi_n$. This limit always exists, because $\{A\psi_n\}$ is a Cauchy sequence if $\{\psi_n\}$ is a Cauchy sequence and A is bounded:

$$\|A\psi_m - A\psi_n\| = \|A(\psi_m - \psi_n)\| \leqslant c\|\psi_m - \psi_n\|.$$

If the domain of A is everywhere dense in $\mathcal{H}$, then the domain of A' is $\mathcal{H}$. For the application of Hilbert space to quantum mechanics, it is sufficient to consider linear operators defined everywhere, or linear operators whose domains are everywhere dense if they are not defined everwhere. Since a bounded linear operator may always be extended uniquely to an operator defined everywhere, if the domain is everywhere dense, the bounded linear operators of quantum mechanics may be considered to be defined everywhere.

A linear operator is *closed* if the convergence of $\{A\psi_n\}$ for a convergent sequence $\{\psi_n\}$ implies that $\lim_{n\to\infty} A\psi_n = A \lim_{n\to\infty} \psi_n$. Notice that equality is required *only in the case that* $\{A\psi_n\}$ *converges*, i.e. when $\lim_{n\to\infty} A\psi_n$ exists. Continuity requires the existence of $\lim_{n\to\infty} A\psi_n$ when $\{\psi_n\}$ converges, and the equality $\lim_{n\to\infty} A\psi_n = A \lim_{n\to\infty} \psi_n$: A is continuous if $\{A\psi_n\}$ converges whenever $\{\psi_n\}$ converges, and $\lim_{n\to\infty} A\psi_n = = A \lim_{n\to\infty} \psi_n$. Every continuous (i.e. bounded) operator may be extended to a closed operator, because a continuous operator may always be extended to an operator defined on a subspace, a closed linear manifold, so that whenever $\lim_{n \to \infty} A\psi_n$ exists, $\lim_{n \to \infty} A\psi_n = A \lim_{n \to \infty} \psi_n$. However, a closed operator is not necessarily continuous – there are unbounded closed linear operators. An unbounded linear operator may be extended to a closed linear operator under very general conditions, and if the closure exists, it is unique. So, although unbounded operators cannot be avoided in quantum mechanics, it is sufficient to consider closed unbounded operators. A closed unbounded operator cannot be defined everywhere: it is a theorem that every closed linear operator on $\mathcal{H}$ is bounded (Jauch, p. 41, after Riesz and Sz.-Nagy). Hence, if a closed linear operator is defined everywhere, it is continuous.

For unbounded operators, it is convenient to distinguish between *self-adjoint* and *Hermitian* operators. The adjoint of an operator A may be defined as the operator A^* satisfying the condition $(A^*\psi, \varphi) = (\psi, A\varphi)$, without assuming that the domains of A^* and A coincide. (Here I follow Jauch's definition rather than von Neumann's.) The domain of A^* is the set of vectors ψ in $\mathcal{H}$ for which the equation $(\psi, A\varphi) = (\psi^*, \varphi)$ is satisfiable by a vector ψ^* in $\mathcal{H}$, i.e. A^* is defined for all such ψ by $A^*\psi = \psi^*$. A is *Hermitian* if A^* is an extension of A, i.e. if the domain of A^* includes the domain of A, and $A^*\psi = A\psi$ for all ψ in the domain of A. A is *self-adjoint* if A^* and A are defined in the same domains and $A^* = A$. In that case, $(A\psi, \varphi) = (\psi, A^*\varphi)$ and $A^{**} = A$. If a closed Hermitian operator is not self-adjoint, i.e. if the domain of A^* contains vectors which do not belong to the domain of A, then generally there are an infinite number of closed Hermitian extensions of A (all of which coincide, of course, only on the domain of A^*). In this set of extensions of A, there may be a subset of operators for which no further extensions are possible, i.e. such operators are already defined at all points where they could be defined without violating their Hermitian character. These operators are said to be *maximal*. A maximal extension which is also self-adjoint is said to be *hypermaximal*.

IV. THE STATISTICAL ALGORITHM

Quantum mechanics incorporates an algorithm for assigning probabilities to ranges of values of the physical magnitudes of a system, and an equation of motion defining the dynamical evolution of the system. The algorithm takes a particularly simple form in the finite-dimensional case.

Consider, firstly, the eigenvalue problem in $\mathcal{H}_n$. This is the problem of finding all solutions to the operator equation

$$A\alpha = a\alpha.$$

If A is a self-adjoint operator, there exist n non-trivial solutions, i.e. n *eigenvectors* α_i and n corresponding *eigenvalues* a_i (besides the trivial solution $\alpha=0$, $a=0$). The eigenvalues are the solutions to the equation

$$|A_{ij} - aI_{ij}| = 0$$

the *secular equation*, where A_{ij} is the $n \times n$ matrix corresponding to the operator A, and $|A_{ij}-aI_{ij}|$ is the determinant of the matrix $A_{ij}-aI_{ij}$, an nth order polynomial in a. (I_{ij} is the unit matrix, with all off-diagonal elements 0 and all diagonal elements 1.) By the fundamental theorem of algebra, the secular equation has n roots. The solutions to the secular equation, the eigenvalues, form the *spectrum* of the operator.

Each a_i is real, since A is self-adjoint:

$$a_i^*(\alpha_i, \alpha_i) = (A\alpha_i, \alpha_i) = (\alpha_i, A\alpha_i) = a_i(\alpha_i, \alpha_i).$$

Eigenvectors corresponding to different eigenvalues are orthogonal, because

$$a_i(\alpha_i, \alpha_j) = (A\alpha_i, \alpha_j) = (\alpha_i, A\alpha_j) = a_j(\alpha_i, \alpha_j)$$

and so $(\alpha_i, \alpha_j)=0$ if $a_i \neq a_j$. Since the solution $\alpha=0$ is excluded, and the vector $a\alpha$ is a solution if α is a solution, it is sufficient to consider solutions α such that $\|\alpha\|=1$, i.e. unit vectors. The eigenvectors of A therefore form an orthonormal set, and since there are exactly n orthogonal eigenvectors, these span $\mathcal{H}_n$, i.e. the orthonormal set is complete.

The n eigenvalues need not all be distinct. If there are $k<n$ *distinct* eigenvalues of A, each eigenvalue $a_i(1 \leqslant i \leqslant k)$ corresponds to $m(i)$ orthogonal eigenvectors $\alpha_{i,1}, \alpha_{i,2}, ..., \alpha_{i,m(i)}$, which span a subspace $\mathcal{H}_{a_i}$ of dimension $m(i)$. The number $m(i)$, the *multiplicity* of the eigenvalue a_i, is the maximum number of linearly independent solutions to the equation $A\alpha=a_i\alpha$. An eigenvalue with multiplicity greater than 1 is said to be *degenerate*. If all the eigenvalues are non-degenerate, the spectrum is *simple*. An operator is degenerate if it has one or more degenerate eigenvalues. (The term 'non-maximal' is sometimes used, and should not be confused with the notion introduced at the end of Section III.)

A degenerate eigenvalue determines a unique subspace $\mathcal{K}_{a_i}$, but since any vector in the subspace $\mathcal{K}_{a_i}$ is a solution to the equation $A\alpha = a_i\alpha$, the eigenvalue a_i does not determine a unique set of $m(i)$ orthogonal vectors in $\mathcal{K}_{a_i}$. (Every vector ψ in $\mathcal{K}_{a_i}$ is expressible in the form

$$\psi = \sum_{j=1}^{m(i)} (\alpha_{i,j}, \psi)\,\alpha_{i,j}$$

so that

$$\begin{aligned}
A\psi &= \sum_{j=1}^{m(i)} (\alpha_{i,j}, \psi)\,A\alpha_{i,j} \\
&= \sum_{j=1}^{m(i)} (\alpha_{i,j}, \psi)\,a_i\alpha_{i,j} \\
&= a_i \sum_{j=1}^{m(i)} (\alpha_{i,j}, \psi)\,\alpha_{i,j} \\
&= a_i\psi\,.
\end{aligned}$$

A complete orthonormal set of eigenvectors, i.e. vector solutions to the eigenvalue equation for a degenerate operator A, is a set of vectors $\{\alpha_{i,j}\}$ where, for each i, $\alpha_{i,1}, \alpha_{i,2}, \ldots, \alpha_{i,m(i)}$ is *any* orthonormal set of vectors spanning $\mathcal{K}_{a_i}$.

In $\mathcal{H}_n$, then, there always exists a complete orthonormal set of eigenvectors for any self-adjoint operator A, i.e. each self-adjoint operator defines a coordinate system or set of basis vectors, which is complete in the sense that any vector ψ in the space is expressible in the form

$$\psi = \sum_{i=1}^{n} (\alpha_i, \psi)\,\alpha_i\,.$$

If A is non-degenerate, the eigenvectors α_i are uniquely determined. If A is degenerate, this is not the case, but it is nevertheless possible to find complete orthonormal sets of solutions to the eigenvalue equation.

Any self-adjoint operator in $\mathcal{H}_n$ may be expressed in the form

$$A = \sum_{i=1}^{k} a_i P_{a_i}$$

where the P_{a_i} are projection operators corresponding to the k orthogonal subspaces $\mathcal{K}_{a_i}$ determined by the k distinct eigenvalues a_i ($\mathcal{K}_{a_i}$ is the subspace of vectors satisfying the equation $A\alpha = a_i\alpha$). This is the *spectral representation* of A. The projection operators P_{a_i} satisfy relations corresponding to the orthogonality and completeness of the eigenvectors:

$$P_{a_i}P_{a_j} = I \quad \text{if} \quad i = j \quad (I \text{ is the unit operator})$$

$$P_{a_i}P_{a_j} = 0 \quad \text{if} \quad i \neq j \quad (0 \text{ is the null operator})$$

$$\sum_{i=1}^{k} P_{a_i} = I.$$

The statistical algorithm has a particularly simple geometric interpretation in $\mathscr{H}_n$. Each self-adjoint operator in $\mathscr{H}_n$ is taken as representing a physical magnitude of the system represented in $\mathscr{H}_n$, in the sense that the spectrum of the operator represents the set of possible values of the magnitude. Notational convention: I use capital Latin letters $A, B, C, \ldots$ to represent physical magnitudes, and I denote the associated Hilbert space operators by the same symbols. I label the eigenvectors of A by the corresponding lower case Greek letter, and the eigenvalues of A by the corresponding lower case Latin letter. In $\mathscr{H}_n$, then, the magnitude A (represented by the operator A) has the possible values $a_1, a_2, \ldots, a_n$ which correspond to the eigenvectors $\alpha_1, \alpha_2, \ldots, \alpha_n$. Each unit vector represents a statistical state, assigning probabilities to ranges of values of the physical magnitudes. If the spectrum of A is simple, i.e. if A has n distinct eigenvalues a_i, the probability assigned to the value a_i of A by the statistical state ψ is

$$p_\psi(a = a_i) = |(\alpha_i, \psi)|^2 = \|P_{a_i}\psi\|^2$$

i.e. $p_\psi(a = a_i)$ is equal to the square of the length of the projection of ψ onto the eigenvector α_i corresponding to a_i.

Here P_{a_i} is the projection operator onto the subspace $\mathscr{H}_{a_i}$, the 1-dimensional subspace spanned by the vector α_i. The expression for the expectation value is

$$\begin{aligned} \text{Exp}_\psi(A) &= \sum_{i=1}^{n} a_i p_\psi(a = a_i) \\ &= \sum_{i=1}^{n} a_i (P_{a_i}\psi, P_{a_i}\psi) \\ &= \sum_{i=1}^{n} a_i (\psi, P_{a_i}\psi) \end{aligned}$$

$$= \left(\psi, \sum_{i=1}^{n} a_i P_{a_i} \psi \right)$$
$$= (\psi, A\psi).$$

(Recall that $A = \sum_{i=1}^{n} a_i P_{a_i}$ in the spectral representation.) Notice that $(\psi, A\psi)$ is always real if A is self-adjoint, because $(\psi, A\psi) = (A\psi, \psi) = (\psi, A\psi)^*$.

This is Born's probabilistic interpretation of the state vector ψ. Schrödinger's time-dependent equation (not to be confused with Schrödinger's eigenvalue equation discussed in Section II) determines a motion of the system represented by a unitary transformation in Hilbert space, i.e. a transformation

$$\psi \to U\psi$$

where U is a unitary operator defined by the condition

$$UU^* = U^*U = I, \quad \text{i.e.} \quad U^* = U^{-1}.$$

It follows that a unitary operator is defined everywhere, continuous (bounded), and preserves the scalar product, and hence the lengths of vectors:

$$(U\psi, U\varphi) = (U^*U\psi, \varphi) = (\psi, \varphi).$$

In the case of a degenerate operator:

$$p_\psi (a = a_i) = \| P_{a_i} \psi \|^2$$

where P_{a_i} is the projection operator onto the $m(i)$-dimensional subspace $\mathscr{K}_{a_i}$. Notice that

$$P_{a_i} = \sum_{j=1}^{m(i)} P_{\alpha_{i,j}}$$

where the $P_{\alpha_{i,j}}$ are the projection operators onto the $m(i)$ mutually orthogonal 1-dimensional subspaces corresponding to *any* orthonormal set of vectors $\{\alpha_{i,j}\}$ spanning $\mathscr{K}_{a_i}$.

Comment on notation: I use the symbol P_s to denote the projection operator onto the subspace identified by the subscript s, which may therefor label the value of a magnitude or a vector (in the case of a 1-dimensional subspace). Thus, P_{a_i} denotes the projection operator onto the

$m(i)$-dimensional subspace associated with the eigenvalue a_i of the magnitude A. $P_{\alpha_i,\,j}$, or in general P_ψ, denotes the projection operator onto the 1-dimensional subspace spanned by the vector. If a_i is a non-degenerate eigenvalue, I use either the symbol P_{a_i} or the symbol P_{α_i} to denote the corresponding subspace, depending on context.

In general, then, in $\mathcal{H}_n$:

$$p_\psi(a \in S) = \sum_{a_i \in S} |(\alpha_i, \psi)|^2 = \sum_{a_i \in S} \|P_{a_i}\psi\|^2 .$$

It is easy to verify that this is indeed a probability assignment to the ranges of values S of A, for a fixed A, where S is any Borel set of real numbers. Since ψ is a vector of unit length, i.e. $\|\psi\|^2 = (\psi, \psi) = 1$, it is obvious geometrically that $0 \leqslant p_\psi(a = a_i) = |(\alpha_i, \psi)|^2 \leqslant 1$, i.e. that the length of the projection of ψ onto a unit vector is less than 1, unless that vector is identical with ψ. And evidently

$$\sum_{i=1}^n p_\psi(a = a_i) = \sum_{i=1}^n |(\alpha_i, \psi)|^2 = 1$$

i.e. the sum of the squares of the lengths of the projections of ψ onto a set of n orthogonal unit vectors is equal to 1, the square of the length of ψ.

Equivalently, from the properties of the scalar product:

$$1 = (\psi, \psi) = \left(\psi, \sum_{i=1}^n (\alpha_i, \psi)\,\alpha_i\right)$$

$$= \sum_{i=1}^n (\alpha_i, \psi)(\psi, \alpha_i)$$

$$= \sum_{i=1}^n (\alpha_i, \psi)(\alpha_i, \psi)^*$$

$$= \sum_{i=1}^n |(\alpha_i, \psi)|^2$$

$$= \sum_{i=1}^n p_\psi(a = a_i).$$

Since each $p_\psi(a = a_i) \geqslant 0$, and $\sum_{i=1}^n p_\psi(a = a_i) = 1$, it follows that $p_\psi(a = a_i) \leqslant 1$, equality corresponding to $\psi = \alpha_i$.

Thus:

(i) $p_\psi(a \in 0) = 0$ $p_\psi(a \in R) = 1$

(ii) $\qquad 0 \leqslant p_\psi\,(a \in S) \leqslant 1$

(iii) $\qquad p_\psi\,(a \in S_1 \quad \text{or} \quad a \in S_2) = p_\psi\,(a \in (S_1 \cup S_2))$
$$= p_\psi\,(a \in S_1) + p_\psi\,(a \in S_2)$$
$$\text{if} \quad S_1 \cap S_2 = 0.$$

In $\mathscr{H}_\infty$ a problem arises because a complete orthonormal set of solutions to the eigenvalue equation $A\alpha = a\alpha$ does not always exist. If this is the case, the spectrum of A is said to be *continuous*. The eigenvalues a_i for which corresponding eigenvectors (or sets of eigenvectors) exist in $\mathscr{H}_\infty$ form the *discrete spectrum* of A. Of course, a countable basis or coordinate system always exists in $\mathscr{H}_\infty$ if $\mathscr{H}_\infty$ is separable. The difficulty here is that the set of eigenvectors of an operator A does not necessarily generate such a coordinate system: it is not always possible to represent an arbitrary vector $\psi \in \mathscr{H}_\infty$ as a linear combination of a countable set of solutions to the eigenvalue equation for A. In this case, it is necessary to reformulate the statistical algorithm of quantum mechanics because the expression

$$p_\psi\,(a \in S) = \sum_{a_i \in S} |(\alpha_i, \psi)|^2$$

is defined only if the eigenvalue equation $A\alpha = a_i\alpha$ has solutions in the range S.

The generalization for $\mathscr{H}_\infty$ is as follows: Each *hypermaximal* operator A defines a unique *spectral measure*, i.e. a map from the field of Borel sets on the real line R into a set of projection operators. Notational convention: The operator A associates the projection operator $P_A(S)$ with the Borel set $S \subseteq R$. The spectral measure satisfies the conditions

$$P_A(0) = 0 \qquad P_A(R) = I$$
$$P_A(S_1 \cap S_2) = P_A(S_1) \wedge P_A(S_2) = P_A(S_1) \cdot P_A(S_2)$$
$$P_A(S_1 \cup S_2) = P_A(S_1) \vee P_A(S_2) = P_A(S_1) + P_A(S_2) - P_A(S_1) \cdot P_A(S_2).$$

Now, each $\psi \in \mathscr{H}_\infty$ assigns a probability $\mu_{\psi A}(S)$ to the range S of A according to the rule

$$p_\psi\,(a \in S) = \mu_{\psi A}(S) = (\psi, P_A(S)\psi) = (P_A(S)\psi, P_A(S)\psi) = \|P_A(S)\psi\|^2.$$

(The symbol $\mu_{\psi A}(S)$ is introduced as an abbreviation for $p_\psi(a \in S)$.) Hence the expectation value of the physical magnitude A in the state ψ is defined by:

$$\text{Exp}_\psi(A) = \int_{-\infty}^{\infty} r \, d\mu_{\psi A}(r)$$

$$= \int_{-\infty}^{\infty} r \, d(\psi, P_A(r) \psi)$$

where $\mu_{\psi A}(r) = \mu_{\psi A}((-\infty, r])$, $P_A(r) = P_A((-\infty, r])$. The operator A may be expressed in the form $A = \int_{-\infty}^{\infty} r \, dP_A(r)$ (the spectral representation of A, cf. $A = \sum_{i=1}^{k} a_i P_{a_i}$ in the discrete case) if we understand $(\psi, \int_{-\infty}^{\infty} r \, dP_A(r)\psi)$ as the integral $\int_{-\infty}^{\infty} d(\psi, P_A(r)\psi)$, so that

$$\text{Exp}_\psi(A) = (\psi, A\psi).$$

Again, one verifies easily that, for a fixed A, $\mu_{\psi A}(S)$ is a probability measure on the set of Borel sets in R. Firstly, $\mu_{\psi A}(S) = (\psi, P_A(S)\psi) = (P_A(S)\psi, P_A(S)\psi) = \|P_A(S)\psi\|^2$, i.e. $\mu_{\psi A}(S)$ is the square of the length of the projection of ψ onto the subspace $\mathcal{H}_A(S)$ corresponding to $P_A(S)$. Since ψ is of unit length, $0 \leqslant \|P_A(S)\psi\|^2 \leqslant 1$. Also, $\mu_{\psi A}(S) = (\psi, P_A(S)\psi) = 0$ if $S = 0$, because $P_A(0) = 0$; and $\mu_{\psi A}(S) = 1$ if $S = R$, because $P_A(R) = I$. If S_1 and S_2 are disjoint sets in R, i.e. $S_1 \cap S_2 = 0$, then

$$\begin{aligned}
\mu_{\psi A}(S_1 \cup S_2) &= (\psi, P_A(S_1 \cup S_2)\psi) \\
&= (\psi, (P_A(S_1) + P_A(S_2))\psi) \\
&= (\psi, P_A(S_1)\psi) + (\psi, P_A(S_2)\psi) \\
&= \mu_{\psi A}(S_1) + \mu_{\psi A}(S_2)
\end{aligned}$$

since $P_A(S_1 \cap S_2) = P_A(S_1) \cdot P_A(S_2) = 0$.

In an n-dimensional Hilbert space, the probability measure $\mu_{\psi A}(r)$ is the Lebesgue-Stieltjes measure concentrated at the eigenvalues a_i ($i = 1, \dots, n$) with weights equal to $p_\psi(a = a_i) = \|P_{a_i}\psi\|^2$. The projection operator $P_A(S)$ is the operator $\sum_{a_i \in S} P_{a_i}$, where P_{a_i} is the projection operator onto the subspace of solutions to the eigenvalue equation corresponding to the eigenvalue a_i. (Comment on notation: My previously introduced symbols P_{a_i}, $\mathcal{H}_{a_i}$ may be regarded as abbreviations for the symbols $P_A(\{a_i\})$, $\mathcal{K}_A(\{a_i\})$, where $\{a_i\}$ denotes the singleton subset containing the eigenvalue a_i.)

The spectral theorem, that each hypermaximal operator on $\mathcal{H}$ defines a unique spectral measure, is evidently a generalization of the theorem that in $\mathcal{H}_n$ the eigenvalue equation is solvable for self-adjoint operators. The solutions to the eigenvalue equation for the self-adjoint operator A in $\mathcal{H}_n$ define a unique spectral measure

$$P_A(S) = \sum_{a_i \in S} P_{a_i}.$$

The algorithm of quantum mechanics for assigning probabilities to ranges of values S of A may be expressed in terms of the spectral measure, without explicit reference to eigenvectors:

$$p_\psi(a \in S) = \| P_A(S)\psi \|^2.$$

Conversely, it can be shown that there are vector solutions $\alpha \neq 0$ to the eigenvalue equation $A\alpha = a\alpha$ in $\mathcal{H}_\infty$ (i.e. in a general $\mathcal{H}$) only at a discontinuity r of $P_A(r)$, and these solutions span a subspace $\mathcal{K}_a$. If the subspaces $\mathcal{K}_a$, for all a, span the Hilbert space, then the eigenvectors of A form a complete orthonormal set. Thus, the points r at which $P_A(r)$ is discontinuous form the discrete spectrum of A, and these are the only values of a for which the eigenvalue equation $A\alpha = a\alpha$, $\alpha \neq 0$, has solutions in $\mathcal{H}$.

The spectrum of A may be defined as the set of points r in whose neighbourhood $P_A(r)$ is not constant. For if $P_A(r)$ is constant for some interval of points $S = [r_1, r_2]$, including r, then

$$\begin{aligned}
S &= (-\infty, r_2] - (-\infty, r_1] \\
&= (-\infty, r_2] \cap (-\infty, r_1]'
\end{aligned}$$

and so

$$\begin{aligned}
P_A(S) &= P_A((-\infty, r_2] \cap (-\infty, r_1]') \\
&= P_A((-\infty, r_2]) \cdot P_A^\perp((-\infty, r_1]) \\
&= P_A(r_2) \cdot (I - P_A(r_1)) \\
&= P_A(r_2) - P_A(r_2) \cdot P_A(r_1) \\
&= P_A(r_2) - P_A(r_1) \\
&= 0.
\end{aligned}$$

(Here $[r_1, r_2]$ denotes a closed interval of points, i.e. an interval containing the end-points r_1, r_2; (r_1, r_2) denotes an interval open on the left, i.e. not containing the end-point r_1.) If $P_A(r)$ is not constant and not discontinous in the neighbourhood of a point r, then the eigenvalue equation has no solutions in Hilbert space for this set of points. Such points form the continuous spectrum of A.

To sum up: The probability assigned to the range S of the magnitude A by the statistical state represented by the vector ψ is defined as

$$p_\psi(a \in S) = \mu_{\psi A}(S) = (\psi, P_A(S)\psi) = \| P_A(S)\psi \|^2.$$

This is the square of the length of the projection of ψ onto the subspace $\mathscr{H}_a(S)$ corresponding to the projection operator $P_A(S)$.

In the case of a finite-dimensional Hilbert space, this expression reduces to

$$p_\psi(a \in S) = \sum_{a_i \in S} |(\alpha_i, \psi)|^2$$

where α_i is the eigenvector corresponding to the eigenvalue a_i of A, i.e. the probability is expressible as the sum of the squares of the projections of ψ onto the eigenvectors corresponding to the eigenvalues lying in the range S.

V. GENERALIZATION OF THE STATISTICAL ALGORITHM

The probability assignments defined by the statistical algorithm for each unit Hilbert space vector ψ may be generalized for convex sets of vectors, i.e. sets of vectors $\psi_1, \psi_2, \ldots$ with weights $w_1, w_2, \ldots$ ($w_1 \geqslant 0, w_2 \geqslant 0, \ldots$; $w_1 + w_2 + \ldots = 1$):

$$p(a \in S) = \sum_i w_i \| P_A(S)\, \psi_i \|^2 = \sum_i w_i (\psi_i, P_A(S)\, \psi_i)$$

and

$$\mathrm{Exp}\,(A) = \sum_i w_i (\psi_i, A\psi_i).$$

It is convenient to express these relations in terms of the *trace* of an operator.

The trace of an operator A is defined as the sum $\sum_i (\varphi_i, A\varphi_i)$, where $\{\varphi_i\}$ is any complete orthonormal set in $\mathscr{H}$. The trace is invariant under a change of basis, i.e.

$$\text{Tr}(A) = \sum_i (\varphi_i, A\varphi_i) = \sum_j (\psi_j, A\psi_j), \quad \text{etc.}$$

If A is self-adjoint, $\text{Tr}(A)$ is real. If A is also definite, (i.e. $(\psi, A\psi) \geqslant 0$ for all ψ), then $\text{Tr}(A) \geqslant 0$, and $\text{Tr}(A)=0$ just in case $A=0$. It can be verified easily that

$$\text{Tr}(aA) = a\,\text{Tr}(A)$$
$$\text{Tr}(A+B) = \text{Tr}(A) + \text{Tr}(B)$$
$$\text{Tr}(AB) = \text{Tr}(BA) \quad \text{for } all \quad A, B \text{ (even non-commuting}$$
$$A, B).$$

If P is a projection operator onto a k-dimensional subspace $\mathscr{K}$, then $\text{Tr}(P)=k$. For

$$\text{Tr}(P) = \sum_{i=1}^{k} (\psi_i, P\psi_i) + \sum_{j=1}^{n} (\varphi_j, P\varphi_j)$$

where $\{\psi_1, \ldots, \psi_k\}$ is any orthonormal set of vectors spanning $\mathscr{K}$, and $\{\psi_1, \ldots, \psi_k; \varphi_1, \ldots, \varphi_n\}$ is an orthonormal set spanning $\mathscr{H}$. Hence

$$\text{Tr}(P) = \sum_{i=1}^{k} (\psi_i, \psi_i) + 0 = k.$$

If $\{\psi_i\}$ is a complete orthonormal set in $\mathscr{H}$, and P_{ψ_i} is the set of projection operators onto the corresponding 1-dimensional subspaces, then

$$\text{Tr}(P_{\psi_j}A) = \text{Tr}(AP_{\psi_j})$$
$$= \sum_i (\psi_i, AP_{\psi_j}\psi_i)$$
$$= (\psi_j, A\psi_j).$$

So, the expectation value of A determined by the convex set of vectors ψ_i with weights w_i is

$$\sum_i w_i \text{Exp}_{\psi_i}(A) = \sum_i w_i(\psi_i, A\psi_i)$$
$$= \sum_i w_i \text{Tr}(P_{\psi_i}A)$$
$$= \text{Tr}\left(\left[\sum_i w_i P_{\psi_i}\right] A\right)$$

The operator $W = \sum_i w_i P_{\psi_i}$, termed the *statistical operator*, is self-adjoint and definite, because each P_{ψ_i} is self-adjoint and definite and $w_i \geqslant 0$. $\mathrm{Tr}(W) = \sum_i w_i = 1$, because $\mathrm{Tr}(P_{\psi_i}) = 1$. The statistical operator completely characterizes the probability assignments defined by convex sets of statistical states according to the rule

$$\mathrm{Exp}_W(A) = \mathrm{Tr}(WA)$$

and

$$p_W(a \in S) = \mathrm{Tr}(WP_A(S))$$

because

$$
\begin{aligned}
p_\psi(a \in S) &= \| P_A(S)\psi \|^2 \\
&= (\psi, P_A(S)\psi) \\
&= \mathrm{Exp}_\psi(P_A(S)) \\
&= \mathrm{Tr}(P_\psi P_A(S)).
\end{aligned}
$$

It is usual to distinguish between *pure statistical states* and *mixtures*, or mixed statistical states. A pure statistical state is determined by a single vector ψ in $\mathcal{H}$ – the corresponding statistical operator is the projection operator P_ψ onto the 1-dimensional subspace $\mathcal{H}_\psi$. Thus, a necessary and sufficient condition that W is the statistical operator of a pure state is that W is a projection operator, or $W^2 = W$ (since W is self-adjoint). If $W^2 \neq W$, W represents a mixture.

The pure statistical states, represented by idempotent statistical operators, are homogeneous in the sense that no idempotent statistical operator is expressible as a convex sum of two (or more) different statistical operators, i.e. if W is idempotent and

$$W = p_1 W_1 + p_2 W_2 \quad (p_1 + p_2 = 1; p_1 > 0, p_2 > 0)$$

then $W = W_1 = W_2$. I reproduce here the simple proof of London and Bauer (pp. 31, 32).

If $W = p_1 W_1 + p_2 W_2$ is idempotent, $W - W^2 = 0$ and

$$
\begin{aligned}
W^2 &= p_1^2 W_1^2 + p_2^2 W_2^2 + p_1 p_2 (W_1 W_2 + W_2 W_1) \\
&= p_1^2 W_1^2 + p_2^2 W_2^2 + p_1 p_2 (W_1^2 + W_1^2 - (W_1 - W_2)^2) \\
&= p_1 W_2^2 + p_2 W_1^2 - p_1 p_2 (W_1 - W_2)^2.
\end{aligned}
$$

So

$$W - W^2 = p_1(W_1 - W_1^2) + p_2(W_2 - W_2^2) + p_1 p_2 (W_1 - W_2)^2 = 0.$$

But

$$(W_1 - W_1^2), (W_2 - W_2^2), (W_1 - W_2)^2$$

are all definite, from which it follows that

$$W_1 - W_1^2 = 0$$
$$W_2 - W_2^2 = 0$$

and, in particular

$$(W_1 - W_2)^2 = 0$$

i.e.

$$W_1 - W_2 = 0$$

since $W_1 - W_2$ is self-adjoint. Thus, $W_1 = W_2$, and from $W_1 = p_1 W_1 + p_2 W_2$, we get $W = W_1 = W_2$.

If W has a pure discrete spectrum of eigenvalues w_i and corresponding eigenvectors ω_i, then W may be expressed in the spectral representation as

$$W = \sum_i \omega_i P_{w_i}.$$

The definiteness of W guarantees that each $w_i \geqslant 0$. Hence, W may be represented as a unique mixture of mutually orthogonal states ω_i with weights w_i. However, if some eigenvalues are degenerate, the set of pure states ω_i is not uniquely determined. What is uniquely determined is the set of distinct eigenvalues with their corresponding subspaces. The subspace $\mathscr{K}_{w_i}$ corresponding to the eigenvalue w_i is the set of all solutions to the eigenvalue equation $W\omega = w_i\omega$. In this case it is not possible to represent W as a unique mixture of orthogonal pure states. In fact, if a mixture is formed from an orthogonal set of k pure states $\psi_1, \psi_2, ..., \psi_k$, each with the same weight (i.e. with the relative weights $1, 1, ..., 1$), then the statistical operator depends only on the subspace spanned by $\psi_1, \psi_2, ..., \psi_k$:

$$W = W'/\mathrm{Tr}(W')$$

where $W' = P_{\psi_1} + P_{\psi_2} + ... + P_{\psi_k} = P_{\mathscr{K}}$. (Note that $\mathrm{Tr}(P_{\mathscr{K}}) = k \neq 1$, the dimension of the subspace $\mathscr{K}$.)

Evidently, the same mixture may be generated by mixing *any* orthonormal set of pure states spanning $\mathscr{K}$ with the relative weights $1, 1, ..., 1$.

Von Neumann's simple example is the mixture obtained from any two orthogonal pure states ψ, φ with equal weights. This yields the same statistics, and is therefore represented by the same statistical operator, as the mixture of pure states $(\psi+\varphi)/\sqrt{2}$, $(\psi-\varphi)/\sqrt{2}$ with equal weights. Thus, while the pure statistical state is pure in the sense that it is homogeneous with respect to the set of statistical states represented by statistical operators in Hilbert space, the mixed state is a mixture only in the sense that it is non-homogeneous – not in the sense that it represents a definite mixture of homogeneous states.

The statistical algorithm of quantum mechanics is now expressed in the general form

$$p_W(a \in S) = \mu_{\psi A}(S) = \mathrm{Tr}(W P_A(S))$$

or

$$\mathrm{Exp}_W(A) = \mathrm{Tr}(WA).$$

This formulation is applicable to both pure and mixed states, finite and infinite dimensional Hilbert spaces, degenerate and non-degenerate operators. A further generalization of this rule to several magnitudes is possible. The generalization is limited to *compatible* magnitudes corresponding to *commuting* operators.

VI. COMPATIBILITY

Two operators A_1 and A_2 are said to *commute* if

$$A_1 A_2 = A_2 A_1.$$

(To avoid complications with different domains of definition, I assume that both A_1 and A_2 are defined everywhere, hence continuous.) If the inverse operators A_1^{-1} and A_2^{-1} exist, then all polynomials of A_1 commute with all polynomials of A_2 (Von Neumann, p. 102). The inverse A^{-1} is a linear operator whose domain is the range of A and whose range is the domain of A, with the property that $A^{-1}A\psi=\psi$ for all ψ in the domain of A, and $AA^{-1}\varphi=\varphi$ for all φ in the range of A (the domain of A^{-1}). The inverse of a linear operator A exists if $A\psi \neq A\varphi$ whenever $\psi \neq \varphi$.

It can be shown that A_1 commutes with A_2 if and only if each $P_{A_1}(S)$ commutes with A_2 and each $P_{A_2}(S)$ commutes with A_1, in fact, if and only if all $P_{A_1}(S)$ commute with all $P_{A_2}(S)$ (Von Neumann, p. 171).

Also, any function of A_1 commutes with any function of A_2 if A_1 and A_2 commute. Conversely, if A_1 and A_2 commute, then there exists a (self-adjoint) operator B such that

$$A_1 = g_1(B)$$
$$A_2 = g_2(B).$$

The operator $g(B)$ is defined as

$$g(B) = \int g(r)\, dP_B(r)$$

where g is a function $g: R \rightarrow R$ (Von Neumann, footnote 94, p. 145). It follows that if $A = g(B)$, then

$$\mu_{WA}(S) = \mu_{WB}(g^{-1}(S)) \quad \text{for every} \quad W, S.$$

This suggests the following definition of *compatibility* for magnitudes: Two physical magnitudes A_1 and A_2 are compatible if and only if there exists a magnitude, B, and functions $g_1: R \rightarrow R$ and $g_2: R \rightarrow R$ such that

$$\mu_{WA_1}(S) = \mu_{WB}(g_1^{-1}(S))$$

and

$$\mu_{WA_2}(S) = \mu_{WB}(g_2^{-1}(S))$$

for all statistical states W and all Borel sets $S \subseteq R$. If $g(B)$ is defined as that magnitude satisfying the relation

$$\mu_{Wg(B)}(S) = \mu_{WB}(g^{-1}(S))$$

for every W and S, where g is again a real-valued function on the real line, then A_1 and A_2 are compatible if and only if $A_1 = g_1(B)$ and $A_2 = g_2(B)$, and the compatibility of the two physical magnitudes is equivalent to the commutativity of the corresponding self-adjoint operators in Hilbert space.

There is an implicit assumption here: that two magnitudes A_1 and A_2 are equivalent if and only if $\mu_{WA_1}(S) = \mu_{WA_2}(S)$ for every W, S. This makes the equivalence classes of quantum mechanical magnitudes, and hence their algebraic structure, depend on the set of statistical states. Different definitions of equivalence are conceivable, especially if the set of statistical states, represented by the statistical operators in Hilbert space, is 'incomplete' in some sense. I shall consider the significance of this equivalence relation again in Chapter VII.

The pure statistical state ψ assigns probabilities to ranges of values S of every magnitude:

$$p_\psi(a \in S) = \|P_A(S)\psi\|^2.$$

In the case of two compatible magnitudes, A_1 and A_2, this expression generalizes to joint probability assignments:

$$p_\psi(a_1 \in S_1 \,\&\, a_2 \in S_2) = \|P_{A_1}(S_1)P_{A_2}(S_2)\psi\|^2.$$

And for n compatible magnitudes $A_1, A_2 \ldots, A_n$:

$$p_\psi(a_1 \in S_1 \,\&\, \ldots \,\&\, a_n \in S_n) = \|P_{A_1}(S_1)\ldots P_{A_n}(S_n)\psi\|^2.$$

(Comment on notation: Here the symbol a_i is a variable denoting a general value of the magnitude A_i, not a name for the ith eigenvalue of a magnitude A as in Section IV. Thus, $a_i \in S_i$ is to be read: The value of the magnitude A_i lies in the range S_i.)

The above relation follows, because if A_1 and A_2 are compatible,

$$A_1 = g_1(B)$$

and

$$A_2 = g_2(B)$$

in the sense that

$$p_\psi(a_1 \in S_1) = p_\psi(b \in g_1^{-1}(S_1))$$

and

$$p_\psi(a_2 \in S_2) = p_\psi(b \in g_2^{-1}(S_2))$$

for every ψ, S_1, S_2. Thus:

$$\begin{aligned}
p_\psi(a_1 \in S_1 \,\&\, a_2 \in S_2) &= p_\psi(b \in g_1^{-1}(S_1) \,\&\, b \in g_2^{-1}(S_2)) \\
&= p_\psi(b \in (g_1^{-1}(S_1) \cap g_2^{-1}(S_2))) \\
&= \|P_B(g_1^{-1}(S_1) \cap g_2^{-1}(S_2))\psi\|^2 \\
&= \|P_B(g_1^{-1}(S_1))P_B(g_2^{-1}(S_2))\psi\|^2 \\
&= \|P_{A_1}(S_1)P_{A_2}(S_2)\psi\|^2.
\end{aligned}$$

In the general case of a mixture represented by the statistical operator W:

$$p_W(a_1 \in S_1 \,\&\, a_2 \in S_2 \,\&\, \ldots \,\&\, a_n \in S_n) = \mathrm{Tr}(WP_{A_1}(S_1)\ldots P_{A_n}(S_n)).$$

Clearly, this expression does not hold for incompatible magnitudes, because

$$\| P_{A_1}(S_1) P_{A_2}(S_2) \psi \|^2 \neq \| P_{A_2}(S_2) P_{A_1}(S_1) \psi \|^2$$

for all ψ, and the order cannot be relevant to the assignment of a joint probability distribution. No further generalization of the statistical algorithm to incompatible magnitudes is possible in Hilbert space.

CHAPTER II

THE PROBLEM OF COMPLETENESS

I. THE CLASSICAL THEORY OF PROBABILITY
AND QUANTUM MECHANICS

The peculiarity of the statistics generated by the algorithm of quantum mechanics can be brought out by recalling certain elementary features of the classical mathematical theory of probability. To introduce terminology, consider an experiment with a finite number of possible outcomes $x_1, x_2 ..., x_n$ of which one and only one can occur. (The experiment of tossing a die, say, has six possible outcomes.) A probability p_i is associated with each outcome x_i, and might, for example, be understood as the relative frequency of the outcome in an infinite (i.e. very long) series of trials or repetitions of the experiment. (In the case of the die, the probabilities are all equal to $\frac{1}{6}$, unles the die is loaded.) An event E is a set of possible outcomes, i.e. a subset of the set $X = \{x_1, x_2, ..., x_n\}$. The probability assigned to the event E is the sum of the probabilities of the outcomes contained in E:

$$p(E) = \sum_{x_i \in E} p_i.$$

(E.g. the probability of an even throw is $p_2 + p_4 + p_6$.) An outcome is an elementary event: $p(\{x_i\}) = p_i$, where $\{x_i\}$ is the singleton subset.

In general, the set X will be infinite, and even uncountable (e.g. if X is the set of possible outcomes of an experiment measuring the position of a free particle). The events E to which probabilities are assigned form a certain set of subsets of X. Evidently, both X (with probability 1) and the null set (with probability 0) belong to the set of events. Also, for every event E, the set of outcomes $E' = X - E$ is an event, the event that occurs if and only if E does not occur. And if E_1, E_2 are events, so is $E_1 \cap E_2$ (the event that occurs if and only if both E_1 and E_2 occur) and $E_1 \cup E_2$ (the event that occurs if and only if either E_1, or E_2, or both occur). The set of events is therefore a field $\mathcal{F}$ of subsets of X (i.e. a non-empty set of subsets closed with respect to finite unions, intersections, and complements), and the probability is a map $p: \mathcal{F} \to R$ satisfying the conditions:

(i) $p(0) = 0;$ $p(X) = 1$
(ii) $0 \leqslant p(E) \leqslant 1$ for $E \in \mathcal{F}$
(iii) $p(E_1 \cup E_2) = p(E_1) + p(E_2)$ if $E_1, E_2 \in \mathcal{F}$ and $E_1 \cap E_2 = 0$.

Condition (ii) is redundant. It follows easily that

$$p(E_1) \leqslant p(E_2) \quad \text{if} \quad E_1 \subseteq E_2$$

since $p(E_1) \leqslant p(E_1) + p(E_2 - E_1) = p(E_2)$, and hence for any E_1, E_2

$$p(E_1 \cup E_2) = p(E_1) + p(E_2 - E_1) \leqslant p(E_1) + p(E_2).$$

(Here $E_2 - E_1 = E_2 \cap E_1'$, where E_1' is the complement of E_1 in X.)

Obviously, since $\mathcal{F}$ is a field, the additivity condition (iii) is equivalent to the condition

$$p\left(\bigcup_{i=1}^{n} E_i\right) = \sum_{i=1}^{n} p(E_i)$$

for every finite class of disjoint sets $E_i \in \mathcal{F}$. For reasons of mathematical convenience, it is usual to extend condition (iii) to countable unions, with the additional assumption that $\mathcal{F}$ is a *σ-field*, i.e. closed under countable unions (and hence closed under countable intersections). The map p is then a normed (i.e. $p(X) = 1$), *σ-additive* (or countably additive) real-valued set function, a probability measure on the σ-field $\mathcal{F}$ of subsets of X. The triple $(X, \mathcal{F}, p)$ is referred to as a probability space.

The axiomatization of the classical theory of probability along these lines is due to Kolmogorov. What is fundamental is the notion of probability as a measure function on an event structure represented as the algebra generated by the subsets of a set under the operations of union, intersection, and complement.

A *random variable* is, loosely, a quantity that can take on different values according to the outcome of an experiment, where the outcomes are the elements of a set X of a probability space $(X, \mathcal{F}, p)$. For example, a quantity which is 1 for even throws of a die and 0 for odd throws is a random variable. More precisely, a random variable is a real-valued function $A:X \to R$. If the probabilities are to be assigned to ranges of values of A, then it is necessary that subsets of X of the form $\{x:a<A(x)<b\}$ belong to $\mathscr{F}$. In general, therefore, it is required that A is a *measurable function* with respect to the σ-field $\mathscr{F}$ of X, i.e. $A^{-1}(S)$ is a *measurable set* – a member of the set $\mathscr{F}$ – for every Borel set $S \subseteq R$. (The class of Borel sets in a topological space is the σ-field generated by the open sets.)

The *expectation* value (average value, mean value) of a random variable A on a probability space $(X, \mathscr{F}, p)$ is defined as

$$\text{Exp}(A) = \int_x A \, dp.$$

Evidently,

$$\text{Exp}(A + B) = \text{Exp}(A) + \text{Exp}(B)$$
$$\text{Exp}(cA) = c \, \text{Exp}(A), \quad c \in R$$

assuming that A and B are random variables on the same probability space and that the integrals all exist. In the special case of a finite probability space with

$$X = x_1, \ldots, x_n, \quad p(\{x_i\}) = p_i, \quad A(x_i) = a_i,$$
$$\text{Exp}(A) = \sum_{i=1}^{n} p_i a_i.$$

A measure of the statistical 'scatter' of A about the expectation value is given by the *variance*, $(\Delta A)^2$, defined as

$$(\Delta A)^2 = \text{Exp}[(A - \text{Exp}(A))^2].$$

Tchebychev's inequality states that if $\text{Exp}(A)$ and A are both finite, then for $\lambda \geqslant 1$

$$p(|A - \text{Exp}(A)| \geqslant \lambda \Delta A) \leqslant \lambda^{-2}.$$

In other words, the probability that A lies outside the interval $(\text{Exp}(A) + - \lambda \Delta A, \text{Exp}(A) + \lambda \Delta A)$ is at most equal to λ^{-2}. If ΔA is small, then the probability that the value of A will be close to $\text{Exp}(A)$ on any particular trial is high. The quantity ΔA is often referred to as the *dispersion* of A. If ΔA is finite, then

$$(\Delta A)^2 = \int_x (A - \mathrm{Exp}(A))^2 \, \mathrm{d}p$$

$$= \int_x A^2 \, \mathrm{d}p - 2 \, \mathrm{Exp} \, A \int_x A \, \mathrm{d}p + (\mathrm{Exp}(A))^2$$

$$= \mathrm{Exp}(A^2) - (\mathrm{Exp}(A))^2.$$

Now, the statistics defined by the algorithm of quantum mechanics is not expressed in terms of the standard classical notion of a probability space. For a single quantum mechanical magnitude, A, the algorithm does define a probability measure in the classical sense, i.e. each statistical state W specifies a measure μ_{WA} on the Borel subsets of the real line. And a set of compatible magnitudes $A_1, ..., A_n$ which are all functions of the magnitude B may be treated as random variables on the probability space for B. The statistics defined for two incompatible magnitudes, however, does not appear to be reducible to the classical scheme: There is no joint probability space for the statistical relations specified by the quantum algorithm for two incompatible magnitudes.

Or is there?

A host of questions arise here: In what sense does the quantum algorithm generate *probabilities*, if the numbers between 0 and 1 specified by the statistical states for ranges of values of the magnitudes do not satisfy Kolmogorov's axioms for probabilities? How are the magnitudes of quantum mechanics to be understood, if they cannot be represented as random variables on a probability space? Or, alternatively, how are the statistical relations to be understood if they do not refer to random variables in the usual sense? Are the statistical states of quantum mechanics 'ultimate' in some sense, or is it possible to introduce additional statistical states which are 'finer' than quantum states, and perhaps additional physical magnitudes, so that the quantum statistics becomes expressible as a truncated part of a classical probability theory? Evidently the relation of compatibility (or incompatibility) is the symptom, and perhaps also the origin, of the non-classical character of the quantum statistics. But what is the significance of this relation?

II. UNCERTAINTY AND COMPLEMENTARITY

Heisenberg's uncertainty principle concerns a reciprocal relationship between the dispersions of certain incompatible magnitudes A_1 and A_2 as the statistical state varies between a dispersion-free state for A_1 and a dispersion-free state for A_2. (A statistical state W is said to be dispersion-free for the magnitude A if $\Delta_W(A) = 0$.) In the case of two incompatible magnitudes satisfying the commutation relation

$$A_1 A_2 - A_2 A_1 = ih/2\pi I$$

e.g. magnitudes corresponding to canonically conjugate classical mechanical quantities such as position and momentum, the relation takes the form

$$\Delta_\psi A_1 \Delta_\psi A_2 \geqslant h/4\pi.$$

The proof is straightforward. Since minimal dispersions are at issue here, only pure states need be considered. First notice that

$$A_1' = A_1 - (\psi, A_1\psi)I$$
$$A_2' = A_2 - (\psi, A_2\psi)I$$

satisfy the same commutation relation as A_1, A_2:

$$A_1'A_2' - A_2'A_1' = ih/2\pi I.$$

Assuming $\|\psi\| = 1$

$$(\psi, (A_1'A_2' - A_2'A_1')\psi) = ih/2\pi(\psi, \psi) = ih/2\pi.$$

Now:

$$\begin{aligned}(\psi, (A_1'A_2' - A_2'A_1')\psi) &= (\psi, A_1'A_2'\psi) - (\psi, A_2'A_1'\psi) \\ &= (A_1'\psi, A_2'\psi) - (A_2'\psi, A_1'\psi) \\ &= 2i \operatorname{Im}(A_1'\psi, A_2'\psi)\end{aligned}$$

where $\operatorname{Im}(A_1'\psi, A_2'\psi)$ is the imaginary part of the complex number $(A_1'\psi, A_2'\psi)$. It follows that

$$\begin{aligned}(\psi, (A_1'A_2' - A_2'A_1')\psi) &\leqslant 2i|(A_1'\psi, A_2'\psi)| \\ &\leqslant 2i\|A_1'\psi\|\|A_2'\psi\|\end{aligned}$$

i.e.

$$\|A_1'\psi\|\|A_2'\psi\| \geqslant h/4\pi.$$

Since

$$(\Delta_\psi A)^2 = (\psi, (A - (\psi, A\psi)I)^2\psi)$$
$$= (\psi, (A')^2\psi)$$
$$= (A'\psi, A'\psi)$$
$$= \|A'\psi\|^2$$

the inequality becomes

$$\Delta_\psi A_1 \Delta_\psi A_2 \geqslant h/4\pi.$$

For Bohr and Heisenberg, who jointly developed the 'Copenhagen interpretation' of the theory, quantum mechanics is irreducibly statistical on the basis of a disturbance theory of measurement, and the uncertainty relations are interpreted as supporting this theory.

A possible reading of Heisenberg's position amounts to something like the following: The measurement procedures by which we determine that the value of a physical magnitude A lies in a certain range S disturb the system in such a way that the values of magnitudes incompatible with A are altered. Furthermore, this disturbance is unavoidable and uncontrollable. A precise analysis of measurement procedures at the microlevel reveals that any procedure for measuring position, for example, involves an interference with the system in such a way that the momentum value is changed in an indeterminate way. (The prototype of this type of argument is the analysis of the γ-ray microscope thought experiment.) The uncertainty relations may be understood as reflecting the *physical* impossibility of simultaneously assigning small ranges of values to incompatible magnitudes in any measurement process, because of the unavoidable and uncontrollable disturbance of the values of magnitudes incompatible with A involved in every measurement of A. The statistical operators of the theory represent those (and only those) probability assignments that are compatible with *our possible knowledge of the microlevel*, in the light of the theory of measurement disturbances. The pure statistical operators represent probability assignments that are compatible with maximal knowledge: they represent various possible totalities of events that are maximal with respect to what can be known simultaneously. These are the state descriptions of quantum mechanics – the quantum analogues of classical states – and the impure statistical operators represent less than maximal knowledge, probability measures over quantum states.

To put this another way: On the basis of a discovery concerning measurement disturbances at the microlevel (i.e. on the assumption that a certain theory of measurement is true), a distinction is made between a maximal totality of events in the sense of classical mechanics, and a totality of events that is maximal with respect to what can be known simultaneously. It is (physically) impossible, by any observation procedure whatsoever, to simultaneously assign ranges of values to a set of magnitudes in such a way as to define a classical state description. The maximal sets of events whose existence and non-existence are simultaneously decidable empirically (at least in principle) are those sets of events corresponding to the quantum state descriptions (i.e. the pure statistical states).

At the risk of introducing a confusing terminology, one might distinguish between a classical event ('c-event') and a quantum event ('q-event'). A c-event is represented by assigning a range to a classical magnitude. A q-event is a c-event whose existence or non-existence can be established by a measurement procedure that does not alter the value of any magnitude entering into the description of the 'q-maximal' totality of events obtaining at that time. Thus, Heisenberg's version of the Copenhagen interpretation might be characterized by the thesis that quantum mechanics is both statistical and complete in the sense that vectors in Hilbert space represent q-maximal totalities of events – totalities of events that are maximal with respect to what can be known simultaneously.

Einstein emphatically rejected this view and proposed a series of arguments designed to show the incompleteness of quantum mechanics. In 1935 he published a paper with Podolsky and Rosen which is directed against Heisenberg's completeness thesis as I have formulated it here. It will be worthwhile to recall the argument, usually referred to as the 'Einstein-Podolsky-Rosen paradox'.

Einstein, Podolsky, and Rosen propose a necessary condition of completeness for a physical theory (Einstein, Podolsky, and Rosen, p. 777):

Every element of the physical reality must have a counterpart in the physical theory.

Secondly, they propose a sufficient criterion of physical reality (Einstein, Podolsky, and Rosen, p. 777):

If, without in any way disturbing a system, we can predict with certainty (i.e. with probability equal to unity) the value of a physical quantity, then there exists an element of physical reality corresponding to this physical quantity.

The condition of completeness, I think, amounts simply to the requirement that every event be representable in the theory, say by assigning a range to a physical magnitude. The criterion of reality is applied in a particular context – that of two separated systems, S_1 and S_2, with correlations between the values of their magnitudes (established during a previous interaction), so that the measurement of an S_1-magnitude makes it possible to assign a value to the correlated S_2-magnitude, without disturbing S_2 by the S_1-measurement. In quantum mechanics, the composite system $S_1 + S_2$ is associated with the tensor product, $\mathscr{H}_1 \times \mathscr{H}_2$, of the Hilbert spaces $\mathscr{H}_1$ of S_1 and $\mathscr{H}_2$ of S_2. (The subscripts here do not denote dimensionality.) The tensor product of $\mathscr{H}_1$ and $\mathscr{H}_2$ is essentially the Cartesian product, $\mathscr{H}_1 \times \mathscr{H}_2$ (which is not, of course, a vector space), with a vector addition operation defined on the set of ordered pairs of vectors, such that

$$\left(\sum_i a_i \psi_i, \sum_j b_j \varphi_j \right) = \sum_{ij} a_i b_j (\psi_i, \varphi_j).$$

Vectors in $\mathscr{H}_1 \otimes \mathscr{H}_2$ are represented by linear combinations of the form $\psi \otimes \varphi$, where $\psi \otimes \varphi$ represents the equivalence class (ψ, φ) under the above equivalence relation. (The symbol (ψ, φ) here denotes an element of $\mathscr{H}_1 \times \mathscr{H}_2$, and not the scalar product.) Operators in $\mathscr{H}_1 \otimes \mathscr{H}_2$ represented as tensor products, $O_1 \otimes O_2$, are defined in the obvious way:

$$O_1 \otimes O_2 \cdot \psi \otimes \varphi = O_1 \psi \otimes O_2 \varphi$$

By an 'S_i-magnitude', I mean a physical magnitude represented by a self-adjoint operator in $\mathscr{H}_i$ ($i = 1, 2$). By an 'S_i-event', I mean an event representable by assigning a range to an S_i-magnitude.

The criterion of reality might be reformulated as follows: In the case of two separated physical systems, S_1 and S_2, an S_i-event is a q-event if a range can be assigned to an appropriate S_i-magnitude as a description of the event by a measurement procedure which does not alter the value of any S_i-magnitude entering into the description of the q-maximal totality of S_i-events obtaining at that time.

The argument proceeds as follows: If the quantum theory is complete (according to the condition of completeness) and the Hilbert space vectors are complete state descriptions in the sense that they represent q-maximal totalities of events, then a position-event (i.e. an event described by assigning a small range of values to the magnitude position) and a momentum event cannot both belong to the q-maximal totality of events obtaining at a particular time. For no q-event corresponds to the joint assignment of a small range to position and a small range to momentum.

Now consider a composite system consisting of two sub-systems, S_1 and S_2, which have interacted suitably at some time so that the values of certain of their magnitudes are correlated. Assume that the Hilbert space vector Ψ of the composite system $S_1 + S_2$ – a vector in the Hilbert space $\mathscr{H} = \mathscr{H}_1 \otimes \mathscr{H}_2$ – is a complete state description in the above sense. In the example considered by Einstein, Podolsky, and Rosen, Ψ does not determine a unique state description for S_1, nor does Ψ determine a unique state description for S_2. What Ψ does determine are certain correlations between S_1-events and S_2-events. These correlations are such that the assignment of a small range to the position of S_1 determines a small range for the position of S_2, and conversely. Similarly, S_1-momentum events are correlated with S_2-momentum events. It follows that whether we assume that a position event belongs to the q-maximal totality of S_1-events obtaining at a particular time, or whether we assume that a momentum event belongs to this totality, we are led to a contradiction. Because if the q-maximal totality of S_1-events includes a position-event, then a small range can be assigned to the momentum of S_1 by a measurement procedure which does not alter the value of any S_1-magnitude entering into the description of that q-maximal totality of S_1-events, viz. a measurement procedure which assigns a small range to the momentum of S_2 (assuming that the two systems are separated, so that there is no possibility of any interaction between them). And, similarly, if the q-maximal totality of S_1-events includes a momentum event, then an S_1-position event is a q-event belonging to that totality.

Thus, the criterion of reality together with the assumption that the Hilbert space vector is a complete state description in Heisenberg's sense leads to a contradiction. In short: If quantum mechanics is complete, then a position event and a momentum event cannot both belong to the q-maximal totality of events obtaining at a particular time. But, in a specific case, it follows from the conjunction of the completeness assumption with the criterion of reality that a position event and a momentum event do both belong to the q-maximal totality of events obtaining at a particular time.

A possible objection to this argument was anticipated by Einstein, Podolsky, and Rosen: One might assume that two (or more) q-events can be asserted to belong to the same q-maximal totality of events only if ranges can be assigned to appropriate magnitudes as a representation of these events by a *simultaneous* process of measurement. Since either an S_1-position-event or an S_1-momentum-event can be established as existing in this way, but not both simultaneously, the argument fails on this supposition. But, as Einstein, Podolsky, and Rosen point out, this would make the q-maximal totality of events obtaining at S_1 at a particular time – the quantum reality at S_1 at a particular time – depend upon the measurement process at S_2, i.e. the kind of measurement being performed at S_2 at that time. Since S_1 and S_2 are separate systems, this appears to be unreasonable.

At this point it might be worthwhile to disentangle Bohr's version of the Copenhagen interpretation from Heisenberg's. Some remarks by Heisenberg on the origin of the Copenhagen interpretation are worthwhile quoting in full (Heisenberg, p. 105; my italics):

At this time, Dirac and Jordan developed the transformation theory to which Born and Jordan in earlier investigations had already laid the foundation, and the completion of this mathematical formalism again confirmed us [i.e. Bohr and Heisenberg] in our belief that there was no more to change in the formal structure of quantum mechanics, and that the remaining problem was to express the connection between the mathematics and the experiment in a way free of contradictions. But how this was to be done remained unclear. Our evening discussions quite often lasted till after midnight, and we occasionally parted somewhat discontented, for the difference in the directions in which we sought the solution seemed often to make the problem more difficult. Still, deeply disquited after one of these discussions I went for a walk in the Faelledpark, which lies behind the Institute, to breathe the fresh air and calm down before going to bed. On this walk under the stars, the obvious idea occurred to me that one should postulate that *nature allowed only experimental situations to occur which could be described within the framework of quantum mechanics.* This would apparently imply, as one could see from the mathematical formalism, that one could not simultaneously know the position and velocity of a particle. There was no immediate possibility of discussing this idea in detail with Bohr, because just at this time (end of February, 1927) he had left for a skiing holiday in Norway. Bohr was probably also glad to be able to devote himself to a few weeks' completely undisturbed thinking about the interpretation of quantum theory.

Left alone in Copenhagen I too was able to give my thoughts freer play, and I decided to make the above uncertainty the central point in the interpretation. Remembering a discussion I had had long before with a fellow student in Göttingen, I got the idea of investigating the possibility of determining the position of a particle with the aid of a gamma-ray microscope, and in this way soon arrived at an interpretation which I believed to be coherent and free of contradictions. I then wrote a long letter to Pauli, more or less the draft of a paper, and Pauli's answer was decidedly positive and encouraging. When Bohr returned from Norway, I was already able to present him with the first version of a paper along with the letter from Pauli. At first Bohr was rather dissatisfied. He pointed out to me that certain statements in this first version were still incorrectly founded, and as he always insisted on relentless clarity in every detail, these points offended him deeply. Further, he had probably already grown familiar, while he was in Norway, with the concept of complementarity which would make it possible to take the dualism between the wave and particle picture as a suitable starting point for an interpretation. *This concept of complementarity fitted well the fundamental philosophical attitude which he had always had, and in which the limitations of our means of expressing ourselves entered as a central philosophical problem.* He therefore took objection to the fact that I had not started from the dualism between particles and waves. After several weeks of discussion, which were not devoid of stress, we soon concluded, not least thanks to Oskar Klein's participation, *that we really meant the same,* and that the uncertainty relations were just a special case of the more general complementarity principle. Thus, I sent my improved paper to the printer and Bohr prepared a detailed publication on complementarity.

How closely the idea of complementarity was in accord with Bohr's older philosophical ideas became apparent through an episode, which, if I remember correctly, took place on a sailing trip from Copenhagen to Svendborg on the island Fyn. At this time Bohr and a colleague and friend owned a sailing boat, the captain of which was the brilliant and extremely charming chemist Bjerrum. The distinguished surgeon Chevitz kept spirits high even in stormy weather, and the other friends contributed each in his own way to this happy and untroubled existence. Bohr was full of the new interpretation of quantum theory, and as the boat took us full sail southwards in sunshine, there was plenty of time to tell of this scientific event and to reflect philosophically on the nature of atomic theory. Bohr began by talking of the difficulties of language, of the limitations of all our means of expressing ourselves, which one had to take into account from the very beginning if one wants to practice science, and he explained how satisfying it was that this limitation had already been expressed in the foundation of atomic theory in a mathematically lucid way. Finally, one of the friends remarked drily, *"But, Niels, this is not really new, you said exactly the same ten years ago."*

One way of understanding the intention behind Heisenberg's remark that "one should postulate that nature allowed only experimental situations to occur which could be described within the framework of the formalism of quantum mechanics" is in terms of the disturbance theory of measurement that I have attributed to Heisenberg in the above discussion of the Einstein-Podolsky-Rosen paradox. It is possible, and perhaps even likely, that Heisenberg's view is more radical than this, along the lines of Bohr's formulation of the Copenhagen interpretation.

Bohr regards the quantum mechanical magnitudes as representing the dispositions of a system to behave (i.e. be 'disturbed') in certain ways in situations defined by macroscopic (classical) systems. The conditions appropriate for the realization of different dispositions may be mutually exclusive. This, Bohr claims, is the case for the space-time and energy-momentum magnitudes of classical physics, i.e. the conditions appropriate for space-time magnitudes exclude the conditions appropriate for energy-momentum magnitudes. Bohr terms pairs of magnitudes (dispositions) which are exclusive in this sense *complementary*, because their simultaneous realization is a presupposition of classical physics: a classical mechanical state is represented by a point in phase space, an assignment of values to position and momentum variables. Quantum mechanics is a rational generalization of classical mechanics in the following sense: Each quantum mechanical magnitude is associated *either* with the group of space-time magnitudes *or* with the group of energy-momentum magnitudes, but not with both. The assignment of a range of values to a magnitude represents an event if and only if the conditions for the realization of the associated disposition are satisfied. The assignment of a probability to a range of values S of a magnitude A is to be understood as the probability that, if the conditions for the realization of the associated disposition were to be satisfied, the corresponding event would obtain.

In effect, Bohr adopts the assumption considered and rejected as unreasonable by Einstein, Podolsky, and Rosen. The contradiction in the application of quantum mechanics to the Einstein-Podolsky-Rosen thought experiment is generated by assuming that the 'reality' at S_1 at a particular time is constituted by *some* q-maximal totality of events obtaining at that time, whether or not we know what these events are. But this assumption is, strictly speaking, incompatible with the quantum description of the composite system, which is represented by a vector Ψ in the Hilbert space $\mathscr{H}_1 \otimes \mathscr{H}_2$ that does not reduce to a product of the form

$$\Psi = \psi_1 \otimes \psi_2$$

where $\psi_1 \in \mathscr{H}, \psi_2 \in \mathscr{H}_2$, nor to a statistical mixture of such products.

In this case, on Bohr's view, no events belonging to the q-maximal totality of events obtaining at a particular time are representable by assigning a range to an S_i-magnitude, because the conditions appropriate for the realization of dispositions associated with S_i-magnitudes are not satisfied (only the conditions appropriate for the realization of dispositions associated with S-magnitudes are satisfied if the composite system is described by the Hilbert space vector Ψ). Moreover, the conditions appropriate for space-time magnitudes at S_2 are simultaneously satisfiable with the conditions for space-time magnitudes at S_1, and exclude the conditions appropriate for energy-momentum magnitudes (and similarly for energy-momentum magnitudes at S_1 and S_2). Thus, an S_1-position measurement (which involves satisfying the conditions appropriate for S_1-space-time magnitudes) excludes the possibility of satisfying the conditions for S_2-momentum-energy magnitudes, and hence excludes the possibility that the assignment of *any* range to the momentum of S_2 represents an S_2-q-event belonging to the q-maximal totality of events obtaining at the time of the S_1-measurement.

To quote from Bohr's reply to Einstein, Podolsky, and Rosen (Bohr (a), p. 700):

From our point of view we now see that the wording of the above-mentioned criterion of physical reality proposed by Einstein, Podolsky, and Rosen contains an ambiguity as regards the meaning of the expression "without in any way disturbing a system." Of course there is in a case like that just considered no question of a mechanical disturbance of the system under investigation during the last critical stage of the measuring procedure. But even at this stage there is essentially the question of *an influence on the very conditions which define the possible types of predictions regarding the future behaviour of the system.* Since these conditions constitute an inherent element of the description of any phenomenon to which the term 'physical reality' can be properly attached, we see that the argumentation of the mentioned authors does not justify their conclusion that quantum-mechanical description is essentially incomplete.

This is the complementarity thesis, that a special mode of description is required for micro-objects, in terms of mutually exclusive groups of dispositions whose simultaneous realization is characteristic of the classical mechanical description of the macrolevel. Thus, a peculiar feature of 'wholeness' is introduced into the description of events by the indivisibility of the micro-object and the measuring instruments which define the conditions under which the events occur. In Heisenberg's terminology, one would have to say that in some sense the measurement process engenders the magnitude measured. Mere disturbance is not enough, as the Einstein-Podolski-Rosen paradox shows. The absurdity of this position was obvious to Bohr (Bohr (b), p. 237):

Meanwhile, the discussion of the epistemological problems in atomic physics attracted as much attention as ever and, in commenting on Einstein's views as regards the incompleteness of the quantum-mechanical mode of description, I entered more directly on questions of terminology. In this connection, I warned especially against phrases, often found in the physical literature, such as "disturbing of phenomena by observation" or "creating physical attributes to atomic objects by measurements". Such phrases, which may serve to remind of the apparent paradoxes in quantum theory, are at the same time apt to cause confusion, since words like 'phenomena' and 'observation', just as 'attributes' and 'measurements', are used in a way hardly compatible with common language and practical definition.

The question is whether complementarity really differs from this *reductio ad absurdam* of Heisenberg's theory of measurement, or whether it amounts to no more than a disguised version of the untenable disturbance theory of measurement. My own view is that Bohr's contribution to the Copenhagen interpretation was that of a remarkably successful propagandist. He saw the statistical relations of quantum mechanics as the confirmation of an approach to the problem of knowledge that had fascinated him since his youth. Heisenberg refers to "the fundamental philosophical attitude which he had always had, and in which the limitations of our means of expressing ourselves entered as a central philosophical problem." Aage Petersen describes this view as "the doctrine that we are, philosophically speaking, suspended in language, that we depend on our conceptual framework for unambiguous communication, and that the scope of the frame may be extended by generalization in the way illustrated in mathematics" (Petersen, p. 10). He quotes Bohr as saying (Petersen, p. 12):

There is no quantum world. There is only an abstract quantum physical description. It is wrong to think that the task of physics is to find out how nature *is*. Physics concerns what we can *say* about nature.

And on another occasion, in reply to an objection that reality is more fundamental than language and 'lies beneath' language, he responded (Petersen, p. 11):

We are suspended in language in such a way that we cannot say what is up and what is down.

Now, one can read almost anything into these intriguing asides, from Plato to Wittgenstein. They reveal Bohr's philosophical hang-ups, no more. The careful phraseology of complementarity, drawing on this reservoir, endows an unacceptable theory of measurement with mystery

and apparent profundity, where clarity would reveal an unsolved problem. Einstein's attitude to the Copenhagen interpretation was less
charitable in private correspondence than would appear from his published articles. In a letter to Schrödinger in May, 1928, he writes:

The Heisenberg-Bohr tranquillizing philosophy – or religion? – is so delicately con-
trived that, for the time being, it provides a gentle pillow for the true believer from
which he cannot very easily be aroused. So let him lie there.

III. HIDDEN VARIABLES

In the 1930's the controversy concerning the completeness of the quantum
statistics focussed on the problem of hidden variables. Since the statistical
operators in Hilbert space do not assign joint probabilities to ranges of
values of incompatible magnitudes, there are no statistical operators
which assign a probability of 1 to particular values of two (or more)
incompatible magnitudes. It follows that there are no statistical operators
which are dispersion-free for all magnitudes.

Now, if the statistical states of a theory are representable in the classical way as probability measures on a σ-field $\mathscr{F}$ of subsets of a set X, the
dispersion-free states correspond to those measures which assign the
value 1 or 0 to every subset in $\mathscr{F}$. To see this, first notice that if a measure
is dispersion-free for every random variable, i.e. every real-valued function on X, then it is dispersion-free for the *characteristic functions* on X.
The characteristic functions represent the idempotent magnitudes of a
classical theory. A characteristic function E maps every point belonging
to some set Y onto 1, and every point outside Y onto 0. Thus:

$$E(x) = 1 \quad \text{if} \quad x \in Y$$
$$E(x) = 0 \quad \text{if} \quad x \notin Y$$

and so

$$E^2 = E.$$

The expectation value of a characteristic function E for a statistical
state W represented by the probability measure p is the probability assigned to the associated set Y by the measure p:

$$\mathrm{Exp}_W(E) = \int_X E \, \mathrm{d}p$$
$$= p(Y).$$

The expression for the dispersion becomes

$$\Delta_W(E) = \mathrm{Exp}_W(E^2) - (\mathrm{Exp}_W(E))^2$$
$$= \mathrm{Exp}_W(E) - (\mathrm{Exp}_W(E))^2$$
$$= p(Y) - p^2(Y).$$

If $\Delta_W(E)=0$, $p(Y)=1$ or 0. Hence, if $\Delta_W(E)=0$ for *every* characteristic function on X, the corresponding probability measure p must satisfy

$$p(Y) = 0 \quad \text{or} \quad 1$$

for *every* set $Y \in \mathscr{F}$.

This is possible only if each dispersion-free probability measure assigns probability 1 to some singleton subset in $\mathscr{F}$. In other words, there is a one-one correspondence between the points of X and dispersion-free probability measures on X, each dispersion-free measure assigning probability 1 to a singleton subset and 0 to the rest of the space. Thus, if the probability assignments of a statistical theory are generated by a set M of statistical states representable as measures on a probability space, the absence of dispersion-free statistical states in M can only indicate the *incompleteness* of the theory, in the sense that the set M does not exhaust *all possible* probability assignments on the event structure represented by the σ-field.

For example, in the case of classical statistical mechanics the statistical states are represented as probability measures on the phase space of classical mechanics. The points of phase space represent classical mechanical states. Each dispersion-free probability measure corresponds to a classical state. If we consider a hypothetical statistical theory involving a set M of statististical states representable as measures on a probability space, with no dispersion-free statistical states in M, we must conclude that the set M is incomplete, since statistical states corresponding to dispersion-free probability measures are definable, and the latter demonstrably exist on the probability space. Relative to the incomplete theory with the set of states M, the variables which parametrize the probability space might well be 'hidden'; they do not belong to the set of physical magnitudes of the M-theory. An assignment of values to these variables would specify a state, analogous to the representation of a classical mechanical state by a point in phase space. But such states would be excluded by the M-theory, since they do not correspond to any assignment of probabilities generated by the statistical states of the set M.

The absence of dispersion-free states in quantum mechanics poses this problem: Is it possible to generate the statistical relations of quantum mechanics by probability measures on a classical probability space? Or is the theory complete in some sense, so that the existence of hidden variables is excluded?

CHAPTER III

VON NEUMANN'S COMPLETENESS PROOF

Unlike Bohr and Heisenberg, who attempted to ground the completeness of quantum mechanics on a thesis concerning the peculiarities of measurement at the microlevel – an argument depending on considerations extraneous to the theory – von Neumann saw the problem as that of proving the non-existence of hidden variables on the basis of certain structural features of quantum mechanics. The difference is important. For suppose quantum mechanics can be reformulated as a classical statistical theory – none of the arguments proposed by Bohr or Heisenberg proves the contrary. Von Neumann could argue that, insofar as quantum mechanics or its completion is an adequate theory, the Copenhagen interpretation is simply false as a thesis concerning micro-elements. But adherents to this interpretation would have to drop quantum mechanics, regretfully perhaps, and search for a new theory incorporating the 'feature of wholeness' or complementarity. The Copenhagen interpretation might conceivably explain why the description of the microlevel requires an 'irreducibly statistical' theory, but it cannot guarantee that the mathematical theory of quantum mechanics does in fact have this character.

Von Neumann opens his investigation by considering the general problem of distinguishing between two alternative interpretations of a statistical theory (Von Neumann, p. 302):

I.   The individual systems $s_1, ..., s_N$ of our ensemble can be in different states, so that the ensemble $[s_1, ..., s_N]$ is defined by their relative frequencies. The fact that we do not obtain sharp values for the physical quantities in this case is caused by our lack of information: we do not know in which state we are measuring, and therefore we cannot predict the results.

II.   All individual systems $s_1, ..., s_N$ are in the same state, but the laws of nature are not causal. Then the cause of the dispersions is not our lack of information, but is nature itself, which has disregarded the 'principle of sufficient cause'.

The question is: Under what conditions is it possible to associate either Case I or Case II *uniquely* with a given statistical theory? Von Neumann's descriptions of the two cases are not formulated very carefully, but the distinction is evidently between a purely epistemic 'ignorance' inter- pretation of the statistics, and an interpretation of the theory as complete, or 'irreducibly statistical' in some sense. Case I characterizes statistical theories in which the probability assignments are generated by statistical states representable as measures on a probability space. Case II is prob- lematic. Is there an objective criterion of demarcation between the two cases, i.e. under what circumstances are we justified in assuming that "the cause of the dispersions is not our lack of information, but is nature itself"? And how are we to understand this notion?

Von Neumann first considers the objection that Case II makes no sense at all. Nature cannot violate the 'principle of sufficient cause', because this amounts to a definition of identity. If two systems behave differently under identical conditions, we would not call them identical. Since measurements of a magnitude A on systems in a statistical ensemble represented by a Hilbert space vector which is not an eigenvector of A yield different results, the systems cannot be identical. That is to say, the Hilbert space description cannot be complete – there must exist other variables, 'hidden parameters', which differentiate the systems in the statistical ensemble represented by the Hilbert space vector.

Now, von Neumann sees a difficulty here because of the supposedly irreducible and uncontrollable interaction between the measuring instru- ment and measured object. Ordinarily, he argues, we might assume that every ensemble can be resolved into homogeneous dispersion-free subensembles, each of which is characterized by particular values for the magnitudes $A, B, C, \ldots$, which are distributed in the original ensemble. But because of the peculiar features of measurement at the microlevel, we cannot actually carry out such a resolution. For suppose we resolve our original ensemble into subensembles characterized by particular values of A. Then any further process of selection of subensembles characterized by particular values of B will change the values of A, if A and B are incompatible (Von Neumann, pp. 304, 305).

That is, we do not get ahead: Each step destroys the results of the preceding one, and no further repetition of successive measurements can bring order into this confusion. In the atom we are at the boundary of the physical world, where each measurement is an interference of the same order of magnitude as the object measured, and therefore affects it basically. Thus the uncertainty relations are at the root of these difficulties.

Therefore we have no method which would make it always possible to resolve further the dispersing ensembles (without change of their elements) or to penetrate to those homogeneous ensembles which no longer have dispersion. The last ones are the ensembles we are accustomed to consider to be composed of individual particles, all identical, and all determined causally. Nevertheless, we could attempt to maintain the fiction that each dispersing ensemble can be divided into two (or more) parts, different from each other and from it, without a change in its elements. That is, the division would be such that the superposition of two resolved ensembles would again produce the original ensemble. As we see, the attempt to interpret causality as an equality definition led to a question of fact which can and must be answered, and which might conceivably be answered negatively. This is the question: Is it really possible to represent each ensemble $[s_1, ..., s_N]$, in which there is a quantity A with dispersion, by a superposition of two (or more) ensembles different from one another and from it?

More precisely: If an ensemble is not dispersion-free, do there always exist two other ensembles such that, for all A:

$$\mathrm{Exp}(A) = c_1 \, \mathrm{Exp}'(A) + c_2 \, \mathrm{Exp}''(A)$$

where c_1 and c_2 are strictly positive and sum to unity? (Here $\mathrm{Exp}'(A)$ and $\mathrm{Exp}''(A)$ are the expectation values of A in the new ensembles, and

$$\mathrm{Exp}(A) \neq \mathrm{Exp}'(A) \neq \mathrm{Exp}''(A).$$

Now, this question can be investigated formally only in the framework of a mathematical theory of probability general enough to incorporate both Case I and Case II statistical theories. For, if we restrict the inquiry to statistical ensembles of the classical (Kolmogorov) type, then it is immediately obvious that all and only dispersion-free ensembles are homogeneous: ensembles which are not dispersion-free are resolvable into homogeneous, dispersion-free ensembles.

Von Neumann recognizes this, and attempts to sketch such a generalized theory implicitly by specifying conditions on the functions $\mathrm{Exp}(A)$, for all A. On the basis of these conditions, von Neumann is able to prove that in the case of quantum mechanics (where the set of magnitudes is represented by the set of self-adjoint operators in Hilbert space, and the functional relationships between magnitudes are those between the corresponding operators) every statistical ensemble is associated with a statistical operator in Hilbert space, which generates the statistics according to the usual rule:

$$\mathrm{Exp}_W(A) = \mathrm{Tr}(WA).$$

The impossibility of generating the statistical relations of quantum mechanics by measures on a classical probability space would now seem to follow because (a) there are no statistical operators representing ensembles which are dispersion-free for *every* magnitude, and yet (b) there *are* homogeneous ensembles, which cannot be resolved into different subensembles. These are represented by the idempotent statistical operators, i.e. projection operators with unit trace.

Von Neumann's theorem is usually misunderstood. The object of the proof is to demonstrate the existence of ensembles which are not dispersion-free and yet homogeneous, given certain conditions on the probability measures and physical magnitudes. It is often pointed out – wrongly, I think – that the proof is redundant: It is sufficient to prove the non-existence of dispersion-free ensembles, the argument goes; the existence of homogeneous ensembles is an added bonus which is not essential to the proof.

Now, von Neumann is obviously not concerned to prove that no quantum mechanical statistical state is dispersion-free, for this follows immediately from the limitation of the statistical algorithm to compatible magnitudes. Nor could the absence of dispersion-free states in the theory by itself exclude the possibility of reconstructing the statistical relations on a classical probability space. Consider, for example, a theory with a set of statistical states corresponding to all possible probability measures on a classical probability space, *excluding the dispersion-free probability measures*. In this case, the theory would include no statistical states representing dispersion-free ensembles, and no statistical states representing homogenous ensembles. Evidently, by construction, the probability assignments of this theory can be generated by measures on a classical probability space.

It is the existence of *homogeneous ensembles with dispersion* that von Neumann regards as significant. His argument is a curious blend of heuristic and logical reasoning here. The absence of dispersion-free statistical states in the theory does not by itself provide a criterion of demarcation between Case I and Case II statistics. Nor, clearly, does the existence of statistical states representing homogeneous ensembles suffice as a criterion. But the uncertainty principle and the Copenhagen theory of measurement suggest that quantum mechanics is irreducibly statistical because ensembles are not resolvable into subensembles with different statistical properties beyond a certain point. In other words, homogeneity with dispersion is the feature characterizing the irreducibly statistical ensembles of micro-objects.

Yet von Neumann's proof does not consist in merely pointing out that quantum mechanics includes such statistical states. Rather, the existence of such states is inferred from general assumptions on the statistical states and physical magnitudes which are presumed to characterize the statistical relations of quantum mechanics. Evidently von Neumann's intention is to develop the proof in the context of a generalized theory of probability. But there is considerable confusion here. The proof fails because von Neumann does not properly distinguish his own foundational problem from the Copenhagen disturbance theory of measurement.

In Chapter V I shall show that von Neumann's criterion of demarcation between Case I and Case II is inadequate. The statistics generated by the algorithm of quantum mechanics on a 2-dimensional Hilbert space satisfies von Neumann's criterion: There are no dispersion-free statistical states, but homogeneous statistical states do exist, viz. those corresponding to idempotent statistical operators or vectors in Hilbert space. It can be shown that these statistical relations *can* be generated by measures on a classical probability space, i.e. the absence of dispersion-free states in the theory, together with the existence of homogeneous statistical states, does *not* exclude the existence of hidden variables here.

Recent criticism of von Neumann's proof has been directed against one particular assumption.

After proposing the following conditions on the functions $\mathrm{Exp}(A)$:

(i)   if the magnitude A is represented by the unit operator, then $\mathrm{Exp}(A)=1$

(ii)   for every A, and every real number a, $\mathrm{Exp}(aA)=a\,\mathrm{Exp}(A)$

(iii) if A is 'by nature' 'non-negative' (e.g. if $A=B^2$), then $\mathrm{Exp}(A)\geqslant0$

(iv)   if A and B are *compatible* magnitudes, then $\mathrm{Exp}(A+B)=\mathrm{Exp}(A)+\mathrm{Exp}(B)$

von Neumann argues that condition (iv) is applicable also to incompatible magnitudes (von Neumann, pp. 308, 309; I have altered von Neumann's symbols to conform to my own notation):

Regarding (iv) it should be noted that its correctness depends on this theorem on probability: the expectation value of a sum is always the sum of the expectation values of the individual terms, independent of whether probability dependencies exist between these or not (in contrast, for example, to the probability of the product). That we have formulated it only for simultaneously measurable A, B, ... is natural, since otherwise $A + B + ...$ is meaningless. But the algorithm of quantum mechanics contains still another operation, which goes beyond the one just discussed: namely, the addition of two arbitrary quantities, which are not necessarily simultaneously observable. This operation depends on the fact that for two Hermitian operators, A, B, the sum $A + B$ is also an Hermitian operator, even if the A, B do not commute, while, for example, the product AB is again Hermitian only in the event of commutativity. In each state the expectation values behave additively:

$$(\varphi, A\varphi) + (\varphi, B\varphi) = (\varphi, (A + B)\varphi) .$$

The same holds for several summands. We now incorporate this fact into our general set-up (at this point not yet specialized to quantum mechanics):

If A, B, ... are arbitrary quantities, then there is an additional quantity $A + B + ...$ (which does not depend on the choice of the $\text{Exp}(A)$-function), such that

$$\text{Exp}(A + B + ...) = \text{Exp}(A) + \text{Exp}(B) + ...$$

J. S. Bell has objected to this extended additivity assumption on the grounds that the eigenvalues of the operator $A + B$ are not the sums of eigenvalues of the operators A and B respectively, if A and B do not commute: measurement of the magnitude $A + B$ is not constituted by adding the results of separate measurements of A and B if these magnitudes are incompatible. In fact, three quite different measurement procedures are involved for A, B, $A + B$, which 'interfere' with each other. Hence, he argues, it is unreasonable to extend the additivity assumption from compatible to incompatible magnitudes. The possibility of generating the probability assignments of quantum mechanics on a classical probability space requires the validity of the additivity assumption only for those probability measures corresponding to the statistical states of quantum mechanics. In particular, the extended additivity assumption might be expected to fail for dispersion-free measures. It seems to me that this argument is confused, although I accept Bell's conclusion that the extended additivity assumption is suspect in the context of the proof. I shall reconsider Bell's argument in Chapter VII and show that it rests on a misunderstanding of the completeness problem.

Doubts such as these about the acceptability of von Neumann's assumptions led to a number of attempts to establish the non-existence of hidden variables on the basis of weaker assumptions. The lattice theoretic proof of Jauch and Piron is perhaps the most successful of these.

CHAPTER IV

LATTICE THEORY: THE JAUCH AND PIRON PROOF

In his *Mathematical Foundations of Quantum Mechanics*, von Neumann pointed out that the properties of a quantum mechanical system are represented by the projection operators in Hilbert space. The properties correspond to the idempotent magnitudes, those magnitudes whose only possible values are 0 and 1, representing the non-existence or existence of the property in question. Since the projection operators are in one-one correspondence with the subspaces of Hilbert space, there is a correspondence between quantum properties, or equivalently quantum *propositions*, and subspaces. Von Neumann remarked (von Neumann, p. 253) that this correspondence "makes possible a sort of logical calculus" with the propositions of quantum mechanics, which differs from ordinary logic in that the notion of 'simultaneous decidability' is relevant. He developed this idea further in a classic paper co-authored with G. Birkhoff. The Birkhoff-von Neumann paper presents the structure of the quantum proposition system as a *lattice* of a certain kind.

A *lattice* is a partially ordered set $\mathscr{L}$ with a greatest lower bound (infimum) and a least upper bound (supremum) defined for every pair of elements. A partially ordered set S is a set with a binary relation $\leqslant$ on S that is reflexive (i.e. $a \leqslant a$), antisymmetric (if $a \leqslant b$ and $b \leqslant a$, then $a = b$), and transitive (if $a \leqslant b$ and $b \leqslant c$ then $a \leqslant c$). The set S is totally ordered if, in addition, either $a \leqslant b$ or $b \leqslant a$ for every $a, b \in S$. The infimum of a and b, denoted by $a \wedge b$, is defined as the (unique) element $c \in \mathscr{L}$ such that $c \leqslant a$ and $c \leqslant b$, and if $c' \leqslant a$ and $c' \leqslant b$, then $c' \leqslant c$. Similarly, the supremum of a and b, denoted by $a \vee b$, is defined as the (unique) element c such that $a \leqslant c$ and $b \leqslant c$, and if $a \leqslant c'$ and $b \leqslant c'$ then $c \leqslant c'$.

It follows that in any lattice:

$$a \wedge b = b \wedge a \qquad\qquad a \vee b = b \vee a$$
$$a \wedge (b \wedge c) = (a \wedge b) \wedge c \qquad a \vee (b \vee c) = (a \vee b) \vee c$$
$$(a \vee b) \wedge b = b \qquad\qquad (a \wedge b) \vee b = b$$

The lower and upper bounds of a lattice are unique of they exist, denoted by 0 and 1, the zero and unit elements. The *complement* of a lattice element a is defined as an element a' such that $a \wedge a' = 0$, $a \vee a' = 1$. The complement a' of a is not necessarily unique. A complemented lattice is a lattice in which every element has a complement. An *ortho-complemented* lattice is a lattice with an operation denoted by $\perp$ (the 'orthogonal complement' or orthocomplement), satisfying the conditions:

$$(a^{\perp})^{\perp} = a$$
$$a \leqslant b \quad \text{if and only if} \quad b^{\perp} \leqslant a^{\perp}$$
$$a \wedge a^{\perp} = 0$$
$$a \vee a^{\perp} = 1$$

The orthocomplement $a^{\perp}$ of a is unique if it exists.

A lattice is *distributive* if

$$a \wedge (b \vee c) = (a \wedge b) \vee (a \wedge c)$$
$$a \vee (b \wedge c) = (a \vee b) \wedge (a \vee c)$$

for every a, b, $c \in \mathscr{L}$. Complementation is unique in a distributive lattice and has the properties of orthocomplementation.

A complemented distributive lattice is a *Boolean algebra*. Indeed, a Boolean algebra may be defined as a non-empty set with two binary operations, meet and join, and one unary operation, complementation, satisfying the axioms (Sikorski, p. 3):

$a \wedge a' = 0$	$a \vee a' = 1$
$a \wedge b = b \wedge a$	$a \vee b = b \vee a$
$a \wedge (b \wedge c) = (a \wedge b) \wedge c$	$a \vee (b \vee c) = (a \vee b) \vee c$
$b \wedge (a \vee b) = b$	$b \vee (a \wedge b) = b$
$a \wedge (b \vee c) = (a \wedge b) \vee (a \wedge c)$	$a \vee (b \wedge c) = (a \vee b) \wedge (a \vee c)$
$0 \vee a = a$	$1 \wedge a = a$

(This set of axioms is redundant.) Essentially, the axioms specify that the three operations have properties analogous to the operations of intersection, union, and complementation on the subsets of a fixed set or space.

The set of subspaces of a Hilbert space forms a lattice under the partial ordering defined by set inclusion. The least upper bound of two subspaces $\mathcal{K}_1$ and $\mathcal{K}_2$ is the smallest subspace containing both $\mathcal{K}_1$ and $\mathcal{K}_2$, and similarly the greatest lower bound of $\mathcal{K}_1$ and $\mathcal{K}_2$ is the largest subspace contained in both $\mathcal{K}_1$ and $\mathcal{K}_2$. The subspace $\mathcal{K}_1 \wedge \mathcal{K}_2$ is identical with the set theoretical intersection of $\mathcal{K}_1$ and $\mathcal{K}_2$, but the union of $\mathcal{K}_1$ and $\mathcal{K}_2$ does not necessarily contain all linear combinations of vectors from $\mathcal{K}_1$ and $\mathcal{K}_2$ (i.e. all vectors of the form $c_1\alpha_1 + c_2\alpha_2$ where $\alpha_1 \in \mathcal{K}_1$ and $\alpha_2 \in \mathcal{K}_2$), and so is not a subspace (and hence not identical with $\mathcal{K}_1 \vee \mathcal{K}_2$).

The lattice of subspaces of a Hilbert space is non-distributive, unlike the Boolean lattice of subsets of a set, i.e. it is not the case in general that

$$\mathcal{K}_1 \wedge (\mathcal{K}_2 \vee \mathcal{K}_3) = (\mathcal{K}_1 \wedge \mathcal{K}_2) \vee (\mathcal{K}_1 \wedge \mathcal{K}_3)$$

or

$$\mathcal{K}_1 \vee (\mathcal{K}_2 \wedge \mathcal{K}_3) = (\mathcal{K}_1 \vee \mathcal{K}_2) \wedge (\mathcal{K}_1 \vee \mathcal{K}_3).$$

Consider, for example, three distinct 1-dimensional subspaces $\mathcal{K}_1$, $\mathcal{K}_2$, $\mathcal{K}_3$ in $\mathcal{H}_2$, such that $\mathcal{K}_2 \perp \mathcal{K}_3$, i.e. $\mathcal{K}_2 = \mathcal{K}_3^\perp$. Then $\mathcal{K}_2 \wedge \mathcal{K}_3 = \mathcal{O}$, the null subspace; so

$$\mathcal{K}_1 \vee (\mathcal{K}_2 \wedge \mathcal{K}_3) = \mathcal{K}_1 \vee \mathcal{O} = \mathcal{K}_1.$$

But

$$\mathcal{K}_1 \vee \mathcal{K}_2 = \mathcal{H}_2$$

and

$$\mathcal{K}_1 \vee \mathcal{K}_3 = \mathcal{H}_2$$

so

$$(\mathcal{K}_1 \vee \mathcal{K}_2) \wedge (\mathcal{K}_1 \vee \mathcal{K}_3) = \mathcal{H}_2.$$

Similarly,

$$\mathcal{K}_2 \vee \mathcal{K}_3 = \mathcal{H}_2,$$

so

$$\mathcal{K}_1 \wedge (\mathcal{K}_2 \vee \mathcal{K}_3) = \mathcal{K}_1 \wedge \mathcal{H}_2 = \mathcal{K}_1$$

whereas

$$\mathcal{K}_1 \wedge \mathcal{K}_2 = \mathcal{K}_1 \wedge \mathcal{K}_3 = \mathcal{O} = (\mathcal{K}_1 \wedge \mathcal{K}_2) \vee (\mathcal{K}_1 \wedge \mathcal{K}_3).$$

A concept of *compatibility*, corresponding to the compatibility of (idempotent) magnitudes, can be defined for subspaces. Two subspaces $\mathcal{K}_1$ and $\mathcal{K}_2$ are compatible if they generate a distributive lattice under the operations of supremum ($\vee$), infimum ($\wedge$), and orthocomplement ($^\perp$). It is easy to see that $\mathcal{K}_1$ and $\mathcal{K}_2$ are compatible if $\mathcal{K}_1 \subseteq \mathcal{K}_2$ or

$\mathcal{K}_1 \subseteq \mathcal{K}_2^\perp$, i.e. if one subspace is included in the other or the two subspaces are orthogonal. It can also be shown that $\mathcal{K}_1$ and $\mathcal{K}_2$ are compatible if and only if they are orthogonal, except for an overlap, i.e. if $\mathcal{K}_1$ and $\mathcal{K}_2$ are expressible in the form

$$\mathcal{K}_1 = [\mathcal{K}_1 \wedge (\mathcal{K}_1 \wedge \mathcal{K}_2)^\perp] \vee (\mathcal{K}_1 \wedge \mathcal{K}_2)$$
$$\mathcal{K}_2 = (\mathcal{K}_1 \wedge \mathcal{K}_2) \vee [\mathcal{K}_2 \wedge (\mathcal{K}_1 \wedge \mathcal{K}_2)^\perp]$$

where

$$\mathcal{K}_1 \wedge (\mathcal{K}_1 \wedge \mathcal{K}_2)^\perp,$$

the intersection of $\mathcal{K}_1$ with the orthogonal complement of $\mathcal{K}_1 \wedge \mathcal{K}_2$, is orthogonal to

$$\mathcal{K}_2 \wedge (\mathcal{K}_1 \wedge \mathcal{K}_2)^\perp,$$

the intersection of $\mathcal{K}_2$ with the orthogonal complement of $\mathcal{K}_1 \wedge \mathcal{K}_2$. (Equivalently, $\mathcal{K}_1$ and $\mathcal{K}_2$ are compatible if there exists three mutually orthogonal subspaces $\mathcal{K}_1'$, $\mathcal{K}_2'$, and $\mathcal{K}$, such that $\mathcal{K}_1 = \mathcal{K}_1' \vee \mathcal{K}$, $\mathcal{K}_2 = \mathcal{K}_2' \vee \mathcal{K}$.) Two subspaces are compatible in this sense if and only if the corresponding projection operators commute, i.e. if and only if the associated idempotent magnitudes are compatible.

The projection operators in $\mathcal{H}$ form a lattice isomorphic to the lattice of subspaces. The partial ordering is defined so that $P_1 \leqslant P_2$ just in case $\mathcal{K}_1 \subseteq \mathcal{K}_2$, where P_1 corresponds to $\mathcal{K}_1$ and P_2 corresponds to $\mathcal{K}_2$ (i.e. $\mathcal{K}_1$ and $\mathcal{K}_2$ are the ranges of P_1 and P_2 respectively). $P^\perp$ is the projection operator $I - P$ with range $\mathcal{H} - \mathcal{K} = \mathcal{K}^\perp$. If $\mathcal{K}_1$ and $\mathcal{K}_2$ are compatible, then

$$P_1 \wedge P_2 = P_1 \cdot P_2$$

and

$$P_1 \vee P_2 = P_1 + P_2 - P_1 \cdot P_2.$$

In general, of course, $P_1 \wedge P_2$ is defined as the projection operator with range $\mathcal{K}_1 \wedge \mathcal{K}_2$, and $P_1 \vee P_2$ as the projection operator with range $\mathcal{K}_1 \vee \mathcal{K}_2$.

If P_1 and P_2 do not commute, i.e. if

$$P_2 P_1 \neq P_2 P_1$$

then $P_1 P_2$ and $P_1 + P_2 - P_1 P_2$ are not projection operators. So, commutativity is a necessary and sufficient conditions for $P_1 P_2$ to be a projection operator. Notice that if $P_1 P_2 = 0$, it follows that $P_2 P_1 = 0$ (because the null operator is a projection operator), and hence that $P_1 + P_2$ is a projection operator if and only if $P_1 P_2 = 0$. Since the subspace corresponding to $P_1 P_2$ is $\mathcal{K}_1 \cap \mathcal{K}_2$ the condition $P_1 P_2 = 0$ is equivalent to the condition that the intersection of $\mathcal{K}_1$ and $\mathcal{K}_2$ is empty, and hence that $\mathcal{K}_1$ is orthogonal to $\mathcal{K}_2$. (Recall that $\mathcal{K}_1$ and $\mathcal{K}_2$ are compatible just in case there exist three mutually orthogonal subspaces $\mathcal{K}'_1$, $\mathcal{K}'_2$, $\mathcal{K} = = \mathcal{K}_1 \cap \mathcal{K}_2$, such that $\mathcal{K}_1 = \mathcal{K}'_1 \vee \mathcal{K}$, $\mathcal{K}_2 = \mathcal{K}'_2 \vee \mathcal{K}$.)

If $P_1, ..., P_n$ is a set of projection operators onto n mutually orthogonal subspaces $\mathcal{K}_1, ..., \mathcal{K}_n$, then $P_1 + P_2 + ... + P_n$ is a projection operator onto the subspace $\mathcal{K}_1 \vee \mathcal{K}_2 \vee ... \vee \mathcal{K}_n$. Also, $P_2 - P_1$ is a projection operator if and only if $P_2 P_1 = P_1$ (or equivalently $P_1 P_2 = P_1$), i.e. if and only if $\mathcal{K}_1 \subseteq \mathcal{K}_2$. In this case $P_2 - P_1$ is the projection operator with range $\mathcal{K}_2 - \mathcal{K}_1 = \mathcal{K}_2 \cap \mathcal{K}_1^\perp$. (Notice that $P_2 - P_1$ is a projection operator if and only if $I - (P_2 - P_1)$ is a projection operator, i.e. if and only if $(I - P_2) + P_1$ is a projection operator. Since a sum of projection operators is a projection operator if and only if their product is the null operator, it follows that $(I - P_2) + P_1$ is a projection operator if and only if $(I - P_2) P_1 = 0$, i.e. $P_1 - P_2 P_1 = 0$.)

Birkhoff and von Neumann's paper was published in 1936, but did not have any appreciable influence on the hidden variable controversy, or the problem of interpretation of quantum mechanics, until the rediscovery of their work by Jauch and co-workers in the 1960's. On the basis of various considerations, which are not of immediate relevance here, Birkhoff and von Neumann proposed that the system of quantum propositions forms a *modular* lattice, i.e. a lattice satisfying the condition

$$a \vee (b \wedge c) = (a \vee b) \wedge c \quad \text{for all} \quad a \leqslant c.$$

Their purpose was to formulate quantum mechanics as a theory on a non-Boolean propositional structure of a certain kind, conceived as a generalization of the classical propositional structure which takes into account measurement restrictions at the microlevel in accordance with Heisenberg's uncertainty principle.

Jauch and Piron find the modularity assumption too restrictive, and develop the theory axiomatically as a generalized probability calculus on a complete, orthocomplemented, weakly modular, atomic lattice.

The assumption of weak modularity is that

$$a \leqslant b \quad \text{only if } a \text{ and } b \text{ are compatible.}$$

In Hilbert space, this axiom is automatically satisfied:

$$\text{If} \quad \mathcal{K}_1 \leqslant \mathcal{K}_2, \quad \mathcal{K}_1 \text{ and } \mathcal{K}_2 \text{ are compatible,}$$

the corresponding projection operators commute, and the associated idempotent magnitudes (properties, propositions) are compatible.

The probability measures on this non-Boolean lattice are defined as generalized probabilities, satisfying the following conditions:

(1) $0 \leqslant p(a) \leqslant 1$ for every $a \in \mathcal{L}$

(2) $p(0) = 0, p(1) = 1$, where 0 and 1 are the minimum ard maximum elements of $\mathcal{L}$

(3)   if $\{a_i\}$ is a sequence of orthogonal elements in $\mathcal{L}$
(i.e. $a_i \leqslant a_k^\perp, i \neq k$) then $p(\vee_i a_i) = \sum_i p(a_i)$

(4)   for any sequence $\{a_i\}$, if $p(a_i) = 1$ for all i then $p(\wedge_i a_i) = 1$

(5a)   if $a \neq 0$, then there exists a probability assignment such that $p(a) = 0$

(5b)   if $a \neq b$, then there exists a probability assignment such that $p(a) \neq p(b)$

The dispersion, Δa, of a proposition a for the probability measure p is defined by

$$\Delta_p^2(a) = \text{Exp}_p(P_a)^2 - (\text{Exp}_p(P_a))^2$$
$$= p(a) - p^2(a)$$

where P_a is the idempotent magnitude or projection operator corresponding to the proposition a. A probability measure is dispersion-free if

$$\Delta_p(a) = 0 \quad \text{for all} \quad a \in \mathcal{L}$$

in which case $p(a) = 1$ or 0 for every $a \in \mathcal{L}$.

Jauch proves the theorem that every generalized probability measure on a lattice is expressible as a weighted integral of dispersion-free measures, only if all the elements of $\mathcal{L}$ are pairwise compatible, i.e. only if $\mathcal{L}$ is Boolean. (This is a modified version of the theorem proved by Jauch and Piron in their joint paper on hidden variables.)

Jauch remarks (Jauch, p. 118):

The conclusion of the theorem is seen to be very strong, since it affirms compatibility for *all* pairs of propositions. Thus it suffices to exhibit a single pair of noncompatible propositions to establish that hidden variables are empirically refuted. Now we have seen that the occurrence of non-compatible propositions is the essence of quantum mechanics, since the lattice is Boolean and the system behaves classically if every pair of propositions is compatible. Because of this result we may simply affirm: a quantum system cannot admit hidden variables in the sense in which we have defined them. With this result the quest for hidden variables of this particular kind has found its definitive answer in the negative.

What the theorem says is this: If the statistical states of a theory are representable as generalized probability measures on a lattice, then these states specify the statistics of ensembles resolvable into homogeneous ensembles defined by dispersion-free states only if the lattice is Boolean.

Now, the possibility of resolving ensembles defined by statistical states on a Boolean lattice into homogeneous ensembles defined by dispersion-free states is no surprise. The classical mathematical theory of probability represents statistical states as measures on a σ-field of subsets of a set, and might equivalently be formalized on a Boolean algebra. Every Boolean algebra is in fact isomorphic to a field of sets. (This isomorphism will be dealt with in detail in Chapter VIII.) Moreover, the impossibility of the resolution in the case of generalized probability measures on a non-Boolean lattice is hardly an interesting piece of information about non-Boolean lattices, but rather a consequence of Jauch's definition of a generalized probability measure.

Consider the set of generalized probability measures on the lattice of subspaces of a 2-dimensional Hilbert space. Every measure corresponding to a quantum mechanical pure state – i.e. associated with a vector in Hilbert space, or a projection operator, according to the statistical algorithm of quantum mechanics – assigns unit probability to *some* proposition (viz. the proposition corresponding to the projection operator, or the 1-dimensional subspace spanned by the vector). Suppose the measure p assigns unit probability to the proposition *a*:

$$p(a) = 1$$

If $\mathcal{K}_a$ and $\mathcal{K}_b$ are incompatible subspaces of $\mathcal{H}_2$, then

$$\mathcal{K}_a \wedge \mathcal{K}_b = \mathcal{O}$$

the null subspace, and so the probability assigned by the measure p to the conjunction of the corresponding incompatible propositions is 0:

$$p(a \wedge b) = 0.$$

By Jauch's condition (4) it follows that

$$p(b) \neq 1$$

for any b incompatible with a. If p' is a dispersion-free measure assigning unit probability to a, then

$$p'(b) = 0$$

for every b incompatible with a. Hence, trivially, the measure cannot be expressed as a weighted integral of dispersion-free measures in Jauch's sense (representing the resolution of the p-ensemble into homogeneous, dispersion-free ensembles), because every single one of these dispersion-free measures necessarily assigns unit probability to a and (by condition (4)) zero probability to b.

Bell has argued that the Jauch and Piron proof, while interesting as a generalization of von Neumann's result to a lattice structure, is open to similar objections with respect to condition (4). All that can reasonably be required of a hidden variable theory is that distributions of hidden variables corresponding to the statistical states of quantum mechanics yield probabilities satisfying condition (4). For particular values of the hidden variables (i.e. for dispersion-free measures, which do not correspond to quantum states), the propositions a and b might both be true, with the proposition $a \wedge b$ false. The proposition $a \wedge b$ is represented by the projection operator $P_{a \wedge b}$ whose range is the subspace $\mathcal{H}_{a \wedge b} = \mathcal{H}_a \wedge \mathcal{H}_b$. Bell's point is that knowledge of the precise values of the hypothetical hidden variables might enable one to predict with certainty that the system will manifest the properties P_a, P_b on measurement, and fail to manifest the property $P_{a \wedge b}$, since the measurement of $P_{a \wedge b}$ does not simply involve a combination of P_a and P_b measurements, but an entirely different measurement procedure, the outcome of which cannot be predicted from a knowledge of measurement results for P_a and P_b.

The evaluation of this criticism raises a number of problems which I have touched on in the previous chapter. Condition (4) is undeniably suspect in the context of a generalized mathematical theory of probability – the most one could require is that the condition holds for compatible lattice elements. But what does *measurement* have to do with the purely internal problem of whether the statistical states of a theory are representable as probability measures on a classical probability space? For that matter, what does the Jauch and Piron theorem have to do with this problem?

Let the statistical states of a theory, say quantum mechanics, be represented as generalized probability measures in Jauch's sense on a non-Boolean lattice. By the theorem, these states cannot be resolved into dispersion-free measures – where a dispersion-free measure is a generalized probability measure in Jauch's sense, satisfying the condition

$$p(a) = 1 \text{ or } 0 \text{ for every } a \in \mathscr{L}.$$

This has absolutely no bearing on the question of whether these states can be represented as probability measures on a classical probability space. The fact that if they are so represented, the states would be resolved into dispersion-free measures does not conflict with the theorem. For the theorem says only that the resolution is impossible in the set of generalized probability measures on the non-Boolean lattice. There might well be no dispersion-free measures in this set with respect to which the resolution can be carried out.

Why, then, has this theorem been proposed as a completeness proof for quantum mechanics? The confusion underlying von Neumann's proof is involved here. The theorem would suffice as a completeness proof for quantum mechanics if it could be guaranteed that the lattice of quantum propositions – the complete, orthocomplemented, weakly modular, atomic lattice – is somehow ultimate, so that the hypothetical Boolean lattice of the classical probability space corresponds to nothing actual. Von Neumann appealed directly to the Copenhagen theory of measurement. Jauch sees the lattice of quantum propositions as "the formalization of a set of *empirical* relations which are obtained by making measurements on a physical system" (Jauch, p. 77). A proposition is a 'yes-no experiment', an experiment with only two possible outcomes, and the non-Boolean features of the quantum lattice are understood as limitations on the possibilities of measurement at the microlevel expressed as empirical relations between yes-no experiments. And this is why "it suffices to exhibit a single pair of noncompatible propositions [interfering yes-no experiments] to establish that hidden variables are empirically refuted" (Jauch, p. 118). Also why Bell's criticism is accepted as relevant: Condition (4) applied to dispersion-free probability measures goes beyond a straight codification of empirical relations between yes-no experiments.

I think this line of reasoning begs the question even more blatantly than von Neumann's argument. And Bell's objection serves only to entrench the confusion. The problem at issue is this: We have before us a mechanics which includes a schema for assigning probabilities to ranges of values of the physical magnitudes. The statistical relations are non-standard and exhibit a number of striking peculiarities. We want to know whether the statistical states, which generate probabilities via an algorithm peculiar to the theory, are representable as measures on a classical probability space. If this is not the case, we want to know just in what sense the numbers generated by the statistical algorithm are probabilities. Since there are no dispersion-free statistical states in the theory, we want to understand the significance of the irreducibility of the statistics. And, lastly, we want to understand the relationship between this mechanics and classical mechanics, which allows a statistical theory of the standard sort.

This problem has been completely solved by Kochen and Specker. In the following Chapter I shall discuss one aspect of their work, their answer to the first of the above questions.

CHAPTER V

THE IMBEDDING THEOREM OF KOCHEN
AND SPECKER

Kochen and Specker begin their analysis of the hidden variable problem by pointing out that in one sense the statistical states of a theory can always be represented as measures on a classical probability space. It is always possible to represent each physical magnitude A by a real-valued function f_A on a space X, i.e. by a random variable on X, and associate each statistical state W with a probability measure ϱ_W on X, so that the measure of the set of points in X mapped onto the set S by the function f_A is equal to the probability assigned to the range S of A by the statistical state W, i.e.

$$\varrho_W(f_A^{-1}(S)) = \mu_{WA}(S)$$

or

$$\int_x f_A(x)\, \mathrm{d}\varrho_W(x) = \mathrm{Exp}_W(A).$$

In the case of quantum mechanics $\mu_{WA}(S) = \mathrm{Tr}(WP_A(S))$ and

$$\mathrm{Exp}_W(A) = \mathrm{Tr}(WA).$$

Consider, for example, a theory with a finite set of physical magnitudes, $A_1, \ldots, A_n$. Introduce a 'hidden variable' x_i for each magnitude, so that the value of x_i determines a value for A_i. Let each hidden variable define a dimension of the probability space X, i.e. the space X is the set of sequences $(x_1, \ldots, x_n)$. Let the value of the function f_{A_i} at the point $(x_1, \ldots, x_n)$ be defined as the value assigned to A_i by x_i. The probability measure corresponding to the statistical state W may be defined as the product measure

$$\varrho_W = \prod_{A_i} \varrho_{WA_i}$$

where

$$\varrho_{WA_i}(f_{A_i}^{-1}(S)) = \mu_{WA_i}(S).$$

Clearly

$$\varrho_W(f_{A_i}^{-1}(S)) = \varrho_{WA_i}(f^{-1}(S)) = \mu_{WA_i}(S)$$

because the set of points in X assigning the range S to the magnitude A_i is the set of sequences $(x_1, ..., x_n)$, where each $x_j \neq x_i$ takes on *all possible values*, and x_i is restricted to the set $f_{A_i}^{-1}(S)$.

In the general case, let X be the set of all possible real valued functions on $\mathcal{Q}$, the set of magnitudes, i.e.

$$X = R^{\mathcal{Q}} = \{x \mid x : \mathcal{Q} \to R\}.$$

Thus, the points of X are functions assigning values to every magnitude, instead of finite sequences. Let the random variable $f_A : X \to R$ be defined as

$$f_A(x) = x(A)$$

and the measure ϱ_W as the product measure

$$\varrho_W = \prod_{A \in \mathcal{Q}} \varrho_{WA}$$

where

$$\varrho_{WA}(f_{A_i}^{-1}(S)) = \mu_{WA}(S).$$

Then:

$$\varrho_W(f_A^{-1}(S)) = \varrho_W(\{x \mid x(A) \in S\}) = \varrho_{WA}(f_A^{-1}(S)) = \mu_{WA}(S).$$

The possibility of this construction, however trivial from a physical point of view, shows that the problem is not the representability of the statistical states of quantum mechanics by measures on a classical probability space. Rather, the problem concerns the possibility of *preserving the structure of the set of physical magnitudes under such a representation*. The trivial construction associates each magnitude with an independent random variable on a probability space, preserving none of the quantum mechanical relations between magnitudes.

The characteristic feature of quantum mechanics is the relation of compatibility, which differentiates the algebraic structure of quantum mechanical magnitudes from the commutative algebra of magnitudes of classical mechanics (or, more generally, from the commutative algebra of random variables on a classical probability space). A set of compatible magnitudes exhibits the structure of a commutative algebra, and might be represented as a set of random variables on a classical probability space, preserving the relationships between magnitudes in the set. But the intransitivity of the compatibility relation disrupts commutativity in a specific way. Kochen and Specker formalize this structure as a *partial algebra*.

A partial algebra over a field $\mathscr{F}$ is a set $\mathscr{A}$ with a reflexive and symmetric binary relation $\leftrightarrow$ (termed 'compatibility'), closed under the operations of addition and multiplication, which are defined only from $\leftrightarrow$ to $\mathscr{A}$, and the operation of scalar multiplication from $\mathscr{F} \times \mathscr{A}$ to $\mathscr{A}$. That is:

(i) $\leftrightarrow \subseteq \mathscr{A} \times \mathscr{A}$

(ii) every element of $\mathscr{A}$ is compatible with itself

(iii) if a is compatible with b, then b is compatible with a, for all $a, b \in \mathscr{A}$

(iv) if any $a, b, c \in \mathscr{A}$ are mutually compatible, then $(a+b) \leftrightarrow c$, $ab \leftrightarrow c$, and $\lambda a \leftrightarrow b$ for all $\lambda \in \mathscr{F}$.

In addition, there is a unit element 1 which is compatible with every element of $\mathscr{A}$, and if a, b, c are mutually compatible, then the values of the polynomials in a, b, c form a commutative algebra over the field $\mathscr{F}$. (Kochen and Specker use the term 'commeasurability' instead of 'compatibility'. For obvious reasons, I prefer the neutral-sounding term.)

A partial algebra over the field $\mathscr{Z}_2$ of two elements, $\{0, 1\}$, is termed a *partial Boolean algebra*. The Boolean operations $\wedge$, $\vee$, and $'$ may be defined in terms of the ring operations $+$, $\cdot$, in the usual way:

$$a \wedge b = a \cdot b$$
$$a \vee b = a + b - a \cdot b$$
$$a' = 1 - a.$$

If a, b, c are mutually compatible, then the values of the polynomials in a, b, c form a Boolean algebra.

Clearly, if $\mathscr{B}$ is a set of mutually compatible elements in a partial algebra $\mathscr{A}$, then $\mathscr{B}$ generates a commutative sub-algebra in $\mathscr{A}$; and in the case of a partial Boolean algebra $\mathscr{A}$, $\mathscr{B}$ generates a Boolean sub-algebra in $\mathscr{A}$. Just as the set of idempotent elements of a commutative algebra forms a Boolean algebra, so the set of idempotents of a partial algebra forms a partial Boolean algebra.

A partial Boolean algebra may be defined directly in terms of the Boolean operations $\wedge$, $\vee$, and $'$. Or, a partial Boolean algebra may be regarded as a partially ordered set with a reflexive and symmetric relation of compatibility, such that each maximal compatible subset is a Boolean algebra. Picture a partial Boolean algebra as 'pasted together' from its maximal Boolean sub-algebras, although the intransitivity of compatibility means that the sub-algebras are not simply pasted side by side and joined at the top and bottom by a common unit (maximum) and zero (minimum) element.

Formally: Take a set of Boolean algebras, $\{\mathcal{B}_i\}_{i \in I}$, such that

(i) the intersection of every pair $\mathcal{B}_i$, $\mathcal{B}_j$ is a Boolean algebra $\mathcal{B}_k = \mathcal{B}_i \cap \mathcal{B}_j$ in the set, and

(ii) if $a_1, ..., a_n$ are elements in $\bigcup_{i \in I} \mathcal{B}_i$, and if for every pair of these elements there exists a Boolean algebra in the set containing both elements of the pair, then there exists a $k \in I$ such that $a_1, ..., a_n \in \mathcal{B}_k$. $\mathcal{B} = \bigcup_{i \in I} \mathcal{B}_i$ is a partial Boolean algebra if the algebraic operations are restricted to elements which lie in a common Boolean algebra $\mathcal{B}_i$, i.e. the compatibility relation is defined by membership in a common $\mathcal{B}_i$.

The problem of representing the statistical states of the quantum algorithm as probability measures on a classical probability space, in such a way that the structure of the set of physical magnitudes is preserved, may be reformulated as the problem of *imbedding the partial algebra of magnitudes, $\mathcal{Q}$, into a commutative algebra*. An *imbedding* of a partial algebra $\mathcal{A}$ into a partial algebra $\mathcal{A}'$ is a homomorphism $h: \mathcal{A} \rightarrow \mathcal{A}'$ which is one-one into $\mathcal{A}'$. A *homomorphism* is a map, $h: \mathcal{A} \rightarrow \mathcal{A}'$, which preserves the algebraic operations, i.e. for all compatible a, $b \in \mathcal{A}$:

$$h(a) \leftrightarrow h(b)$$
$$h(\lambda a + \mu b) = \lambda h(a) + \mu h(b)$$
$$h(ab) = h(a) h(b)$$
$$h(1) = 1.$$

Now, the association

$$A \rightarrow f_A$$

represents the magnitudes as random variables $f_A: X \rightarrow R$ on a probability space X, which form a commutative algebra $\mathcal{C}$. If the representation of the quantum statistics on a classical probability space is to preserve the structure of the set of physical magnitudes, $\mathcal{Q}$, the association $A \rightarrow f_A$ must be an imbedding of $\mathcal{Q}$ into $\mathcal{C}$.

The imbeddability of the partial algebra of magnitudes into a commutative algebra requires the imbeddability of the partial Boolean algebra of idempotent magnitudes into the set of idempotents of the commutative algebra, i.e. into a Boolean algebra. Kochen and Specker prove the preliminary theorem that a necessary and sufficient condition for the imbeddability of a partial Boolean algebra, $\mathscr{A}$, into a Boolean algebra is that for every pair of distinct elements a, $b \in \mathscr{A}$ there exists a 2-valued homomorphism, $h: \mathscr{A} \to \mathscr{Z}_2$, separating these elements, i.e. such that

$$h(a) \neq h(b).$$

Notice that each point x of the probability space X assigns a definite value $f_A(x)$ to every magnitude A. This assignment is actually a real-valued homomorphism on the partial algebra of magnitudes, and a 2-valued homomorphism on the partial Boolean algebra of idemptotents, the quantum propositions.

Kochen and Specker prove that the partial Boolean algebra of idempotent magnitudes on a 3-dimensional Hilbert space cannot be imbedded into a Boolean algebra by showing that there are no 2-valued homomorphisms on the partial Boolean algebra of subspaces of $\mathscr{H}_3$.

This is not difficult to see. The following proof is an adaptation of a similar proof by Bell. (Bell's proof is a demonstration that in a Hilbert space of 3 or more dimensions von Neumann's additivity condition cannot be satisfied by dispersion-free statistical states, not even for the expectation values of compatible magnitudes.)

If $\mathscr{H}_1$, $\mathscr{H}_2$, $\mathscr{H}_3$ are mutually orthogonal 1-dimensional subspaces of $\mathscr{H}_3$, and h is a 2-valued homomorphism, then:

$$h(\mathscr{H}_1) \vee h(\mathscr{H}_2) \vee h(\mathscr{H}_3) = h(\mathscr{H}_1 \vee \mathscr{H}_2 \vee \mathscr{H}_3) = h(\mathscr{H}_3) = 1$$

$$h(\mathscr{H}_i) \wedge h(\mathscr{H}_j) = h(\mathscr{H}_i \wedge \mathscr{H}_j) = h(\mathcal{O}) = 0 \ (i, j = 1, 2, 3; i \neq j).$$

Hence, h maps one and only one of every orthogonal triple of lines onto 1, the remaining two lines being mapped onto 0. If the lines are replaced by lines of unit length, then h defines a map from the surface of the unit sphere onto $\{0, 1\}$, such that for any orthogonal triple of points on S, exactly one point is mapped onto 1. It follows that if two points on S, represented by two unit vectors, α and β, are orthogonal, and $h(\alpha) = = h(\beta) = 0$, then $h(a\alpha + b\beta) = 0$, for all a, b, i.e. any vector in the plane spanned by α, β is assigned the value 0.

Now consider any pair of unit vectors, α and β, such that $v(\alpha) = 1$ and

$v(\beta)=0$. The vector β may be expressed as a linear combination

$$\beta = \alpha + p\alpha'$$

where α' is a unit vector orthogonal to α. Let α'' be the remaining member of the orthogonal triple of unit vectors $\{\alpha, \alpha', \alpha''\}$.

If it were possible to demonstrate that

$$v(\alpha + \alpha') = v(\alpha - \alpha') = 0$$

then it would follow that $v(a\alpha)=0$, since $\alpha+\alpha''$ and $\alpha-\alpha''$ are orthogonal and sum to 2α. But $v(2\alpha)=v(\alpha)$, which was assumed equal to 1. Now, it would follow that $v(\alpha+\alpha'')=0$ if $\alpha+\alpha''=\gamma+\delta$, where γ is a linear combination of β and α'', and δ is a linear combination of α' and α'', and γ and δ are orthogonal. For this to be possible, it is sufficient that

$$\gamma = \beta + q^{-1}p\alpha'' = \alpha + p\alpha' + q^{-1}p\alpha''$$

and

$$\delta = - p\alpha' + qp\alpha''$$

with

$$p(q + q^{-1}) = 1, \quad \text{i.e.} \quad p \leqslant \tfrac{1}{2}.$$

Similarly, if $p \leqslant \tfrac{1}{2}$ there are real values of q such that $p(q+q^{-1})=-1$, i.e. such that $\alpha-\alpha''=\gamma+\delta$. Thus a contradiction follows unless $p>\tfrac{1}{2}$, i.e. unless there is a minimum distance between vectors assigned different values.

Now, since each vector is assigned either a 1 or a 0, there must be pairs of arbitrarily close vectors that are assigned different values, i.e. there cannot be a minimum distance between vectors with different values. For suppose ψ and φ are two points on S that are assigned different values. Then, no matter how close ψ and φ are on S, there exists a point χ between ψ and φ. Since χ is assigned either a 1 or a 0, the value assigned to χ must either be the same as the value assigned to ψ or the same as the value assigned to φ. (This is a topological property of the surface S.) Thus, a contradiction cannot be avoided on the assumption that the map exists.

The proof of Kochen and Specker is rather more complicated. They consider it important to show that there is no 2-valued homomorphism on a *finite* partial Boolean sub-algebra of the partial Boolean algebra of subspaces of $\mathcal{H}_3$. Thus, their proof establishes the non-existence of a map from a *subset*, T, on the surface of the unit sphere onto $\{0, 1\}$, such that for any orthogonal triple of points in T, exactly one is mapped onto 1. Actually, they require 117 points on the sphere with particular orthogonality relations. They show that there is no way of assigning 1's and 0's to this set of points in such a way that exactly one member of every orthogonal triple is mapped onto 1. The above proof does not construct a finite set of points on the surface of the sphere which can be mapped onto $\{0, 1\}$ only by violating the orthogonality conditions. What is shown instead is that the orthogonality conditions can only be satisfied if there is a minimum distance between points mapped onto different values. And this conflicts with the topology of the surface of the sphere.

The proof applies to Hilbert spaces of three or more dimensions. In the case of $\mathscr{H}_2$, the unit vectors define the circumference of the unit circle, and the mapping is possible. For example, if the points are labelled by an angular parameter θ between 0 and 2π, a 2-valued homomorphism on the partial Boolean algebra of subspaces of $\mathscr{H}_2$ is defined by a map which assigns 1 to all points in the half-open quadrants $[0, \pi/2)$, $[\pi, 3\pi/2)$, and 0 to all points in the half-open quadrants $[\pi/2, \pi)$, $[3\pi/2, 2\pi)$. This difference between a quantum mechanical system associated with a 2-dimensional Hilbert space, and quantum mechanical systems associated with higher dimensional Hilbert spaces is crucially important for the interpretation of quantum mechanics. For the imbeddability of the partial Boolean algebra of subspaces of $\mathscr{H}_2$ into a Boolean algebra means that it is possible to represent the statistical relations generated by the quantum algorithm on $\mathscr{H}_2$ by measures on a classical probability space, in such a way that the algebraic structure of the set of magnitudes is preserved. In other words, a hidden variable reconstruction of the quantum statistics is possible in this case, and Kochen and Specker formulate such a theory.

Notice that the uncertainty principle holds for $\mathscr{H}_2$, i.e. the uncertainty relations (and hence the disturbance theory of measurement) cannot characterize the irreducibility of the quantum statistics, because they hold even when the statistics is formally reducible.

Putting this another way: There are no dispersion-free statistical states on the partial Boolean algebra of subspaces of $\mathscr{H}_2$, but homogeneous statistical states do exist. Thus, von Neumann's criterion of demarcation between 'classical' statistical theories (Case 1) and irreducibly statistical theories (Case II) is inadequate.

CHAPTER VI

THE BELL-WIGNER LOCALITY ARGUMENT

J. S. Bell has objected to the Kochen and Specker imbedding theorem as a proof of the completeness of quantum mechanics on similar grounds to his objections to von Neumann's proof and the proof of Jauch and Piron. Again, the argument is that the proof implicitly assumes that equal values are assigned to quantum mechanically equivalent magnitudes by the hypothetical dispersion-free probability measures, whereas equivalence in the algebra of quantum mechanical magnitudes should be understood as statistical equivalence only for those probability measures which correspond to the statistical states of quantum mechanics.

Bell proposes a 'locality condition' as a physically motivated restriction on hidden variable theories which attempt to reconstruct the quantum statistics on a classical probability space. In the case of separated but coupled systems as in the Einstein-Podolski-Rosen experiment, he derives an inequality for certain statistical relations on the classical probability space which the quantum mechanical probabilities fail to satisfy. On this basis local hidden variable theories are rejected.

Before discussing Bell's objection to the completeness proofs of von Neumann, Jauch and Piron, and Kochen and Specker in Chapter VII, I want to show that Bell's result is quite trivial and irrelevant to the completeness problem. I shall develop Wigner's elegant reformulation of Bell's argument.

Wigner begins by considering something like the hidden variable construction outlined at the beginning of Chapter V for a system with a finite set of physical magnitudes. The system in question is the quantum mechanical system consisting of two spin-$\frac{1}{2}$ particles in the singlet spin state, and the magnitudes are the nine magnitudes corresponding to spin in any of three directions **a**, **b**, **c** for the two particles.

The quantum mechanical description of the composite system is as follows: Each subsystem is associated with a 2-dimensional Hilbert space. The magnitude 'spin in the direction **a**' is represented by a self-adjoint operator A on the Hilbert space, with two distinct eigenvalues,

a_+ and a_-, and corresponding eigenvectors, α_+ and α_-. The eigenvalues a_+, a_- correspond to 'spin up in the direction **a**', and 'spin down in the direction **a**', respectively, in the usual terminology. Similarly, the magnitudes 'spin in the direction **b**', and 'spin in the direction **c**' are represented by the operators B, C with eigenvalues b_+, b_-; c_+, c_-, and corresponding eigenvectors β_+, β_-; γ_+, γ_-.

I shall use superscripts to distinguish the two subsystems and their descriptions. The composite system, $S^1 + S^2$, is associated with the tensor product Hilbert space, $\mathcal{H}^1 \otimes \mathcal{H}^2$. Recall that the tensor product of $\mathcal{H}^1$ and $\mathcal{H}^2$ is essentially the Cartesian product, $\mathcal{H}^1 \times \mathcal{H}^2$ (which is not a vector space), with a vector addition operation defined on the set of ordered pairs of vectors, such that

$$\left(\sum_i k_i^1 \psi_i^1, \sum_j k_j^2 \psi_j^2 \right) = \sum_{ij} k_i^1 k_j^2 \, (\psi_i^1, \psi_j^2)$$

(where the symbol (ψ^1, ψ^2) here denotes an element of $\mathcal{H}^1 \otimes \mathcal{H}^2$, and not the scalar product).

Vectors in $\mathcal{H}^1 \otimes \mathcal{H}^2$ are represented by linear combinations of the form $\psi^1 \otimes \psi^2$, where $\psi^1 \otimes \psi^2$ represents the equivalence class (ψ^1, ψ^2) under the above equivalence relation. Operators in $\mathcal{H}^1 \otimes \mathcal{H}^2$ represented as tensor products, $O^1 \otimes O^2$, are defined in the obvious way:

$$O^1 \otimes O^2 \cdot \psi^1 \otimes \psi^2 = O^1 \psi^1 \otimes O^2 \psi^2 .$$

The nine magnitudes are those represented by the operators

$$
\begin{array}{lll}
A^1 \otimes A^2, & A^1 \otimes B^2, & A^1 \otimes C^2 \\
B^1 \otimes A^2, & B^1 \otimes B^2, & B^1 \otimes C^2 \\
C^1 \otimes A^2, & C^1 \otimes B^2, & C^1 \otimes C^2 .
\end{array}
$$

These magnitudes have four eigenvalues, corresponding to the spin values $++$, $+-$, $-+$, $--$. In the case of the magnitude $A^1 \otimes B^2$, the eigenvectors are

$$\alpha_+^1 \otimes \beta_+^2, \quad \alpha_+^1 \otimes \beta_-^2, \quad \alpha_-^1 \otimes \beta_-^2, \quad \alpha_-^1 \otimes \beta_-^2 .$$

The singlet spin state is the statistical state represented by the vector

$$\Psi = (1/\sqrt{2})\, \alpha_+^1 \otimes \alpha_-^2 - (1/\sqrt{2})\, \alpha_-^1 \otimes \alpha_+^2$$

where Ψ has been expressed as a linear combination of the basis vectors

$$\alpha_+^1 \otimes \alpha_+^2, \quad \alpha_+^1 \otimes \alpha_-^2, \quad \alpha_-^1 \otimes \alpha_+^2, \quad \alpha_-^1 \otimes \alpha_-^2 .$$

In terms of the β-system, or γ-system:

$$\Psi = (1/\sqrt{2})\,\beta_+^1 \otimes \beta_-^2 - (1/\sqrt{2})\,\beta_-^1 \otimes \beta_+^2$$
$$= (1/\sqrt{2})\,\gamma_+^1 \otimes \gamma_-^2 - (1/\sqrt{2})\,\gamma_-^1 \otimes \gamma_+^2 \,.$$

The statistical correlations defined by Ψ are such that

$$p(a^1 = a_+^1 \ \& \ a^2 = a_-^2) = \tfrac{1}{2}$$
$$p(a^1 = a_-^1 \ \& \ a^2 = a_+^2) = \tfrac{1}{2}$$
$$p(a^1 = a_+^1 \ \& \ a^2 = a_+^2) = 0$$
$$p(a^1 = a_-^1 \ \& \ a^2 = a_-^2) = 0$$

i.e. S^1 and S^2 never have a^1 and a^2 both positive ('spin in direction **a** up'), or both negative ('spin in direction **a** down'): if a^1 is positive, a^2 is negative, and conversely. The statistical correlations for b and c are similar. Thus, S^1 and S^2 are 'mirror-images' of each other for the magnitudes A, B, C.

Now, the classical probability space X is the space of sequences $(x_1, \ldots, x_9)$, where each hidden variable determines a value of the corresponding magnitude (assuming some ordering of the nine magnitudes). Thus, we have a 9-dimensional Cartesian space, which can be partitioned into 4^9 subsets corresponding to the 4^9 distinct sequences of values for the 9 magnitudes.

Bell's locality condition is this: In the case of measurements on two separated systems S^1 and S^2 the *result* of a measurement on S^1 cannot be affected by the *kind* of measurement performed on S^2 (e.g. whether it is a measurement of spin in the direction **a**, i.e. the magnitude A^2, or whether it is a measurement of spin in the direction **b**, i.e. the magnitude B^2). That is to say, the value of an S^1-magnitude A^1 should depend only on the values of the hidden variables (and perhaps the experimental arrangement, according to Bohr) at S^1 and not at all on the experimental arrangement at S^2, if S^1 and S^2 are sufficiently separated in space.

The locality condition reduces the 9 distinct magnitudes to 6: measurement of A^1, for example, is no longer regarded as a partial measurement of either $A^1 \otimes A^2$ or $A^1 \otimes B^2$ or $A^1 \otimes C^2$. The 6 distinct magnitudes are

$$A^1, B^1, C^1$$
$$A^2, B^2, C^2$$

each with 2 possible values:

$$a^1_+, a^1_- : \quad b^1_+, b^1_- ; \quad c^1_+, c^1_-$$
$$a^2_+, a^2_- ; \quad b^2_+, b^2_- ; \quad c^2_+, c^2_-$$

Only 6 hidden variables are introduced, i.e. a 6-dimensional probability space which can be partitioned into 2^6 subsets corresponding to the 2^6 distinct sequences of values for the 6 magnitudes.

A brief argument now shows that no probability measure on this space can reproduce the quantum mechanical probabilities generated by the singlet spin state. Let $(i, j, k; l, m, n)$ denote the measure of the subset in the probability space corresponding to the sequence of values $(a^1_i, b^1_j, c^1_k;$ $a^2_l, b^2_m, c^2_n)$ for the magnitudes, where the subscripts range over $+$ and $-$. In the case of the singlet spin state, all but 8 of the 2^6 subsets have measure zero, for $(i, j, k; l, m, n)=0$ if $i=l$ or $j=m$ or $k=n$.

It follows that

$$p(a^1 = a^1_+ \,\&\, c^2 = c^2_+) = \sum_{jklm} (+, j, k; l, m, +)$$
$$= (+, +, -; -, -, +) + (+, -, -; -, +, +)$$
$$= w + z, \quad \text{say}$$
$$p(b^1 = b^1_+ \,\&\, c^2 = c^2_+) = (+, +, -; -, -, +) + (-, +, -; +, -, +)$$
$$= w + x, \quad \text{say}$$
$$p(a^1 = a^1_+ \,\&\, b^2 = b^2_+) = (+, -, +; -, +, -) + (+, -, -; -, +, +)$$
$$= y + z, \quad \text{say}.$$

Since w, x, y, z are all positive and greater than or equal to zero:

$$p(a^1 = a^1_+ \,\&\, c^2 = c^2_+) \leqslant p(b^1 = b^1_+ \,\&\, c^2 = c^2_+) + p(a^1 = a^1_+ \,\&\, b^2 = b^2_+).$$

The probabilities generated by Ψ are:

$$p(a^1 = a^1_+ \,\&\, c^2 = c^2_+) = \tfrac{1}{2} \sin^2 \tfrac{1}{2}\theta_{ac}$$
$$p(b^1 = b^1_+ \,\&\, c^2_2 = c^2_+) = \tfrac{1}{2} \sin^2 \tfrac{1}{2}\theta_{bc}$$
$$p(a^1 = a^1_+ \,\&\, b^2 = b^2_+) = \tfrac{1}{2} \sin^2 \tfrac{1}{2}\theta_{ab}$$

where θ_{ac} is the angle between **a** and **c**, etc. But the inequality

$$\sin^2 \tfrac{1}{2}\theta_{ac} \leqslant \sin^2 \tfrac{1}{2}\theta_{ab} + \sin^2 \tfrac{1}{2}\theta_{bc}$$

cannot be satisfied for arbitrary angles. For example, if **b** bisects the angle between **a** and **c**, the inequality does not hold, and it is easy to demonstrate that the inequality fails to hold whenever **a**, **b**, and **c** are coplanar.

It would seem that the possibility of representing the statistical states of quantum mechanics by measures on a classical probability space is excluded on the basis of a simple calculation, at least in the case of representations satisfying the locality condition. This makes the imbedding theorem of Kochen and Specker uninteresting.

Now, a similar inequality may be derived for a single spin-$\frac{1}{2}$ particle, associated with a 2-dimensional Hilbert space. Consider the three magnitudes A, B, C: spin in the direction $\mathbf{a}$, spin in the direction $\mathbf{b}$, spin in the direction $\mathbf{c}$. The probability space is 3-dimensional, partitioned into eight subsets corresponding to the $2^3 = 8$ distinct sequences of values for the 3 magnitudes. We have:

$$p(a = a_- \ \& \ c = c_+) = (-, +, +) + (-, -, +)$$
$$p(b = b_- \ \& \ c = c_+) = (+, -, +) + (-, -, +)$$
$$p(a = a_- \ \& \ b = b_+) = (-, +, +) + (-, +, -).$$

It follows that

$$p(a = a_- \ \& \ c = c_+) \leqslant p(a = a_- \ \& \ b = b_+) + p(b = b - \ \& \ c = c_+).$$

The pairs of eigenvectors

$$\alpha_+, \alpha_- ; \qquad \beta_+, \beta_- ; \qquad \gamma_+, \gamma_-$$

in Hilbert space are related in such a way that

$$\alpha_+ = (1/\sqrt{2}) \cos \tfrac{1}{2}\theta_{ab}\beta_+ + (1/\sqrt{2}) \sin \tfrac{1}{2}\theta_{ab}\beta_-$$
$$\alpha_- = -(1/\sqrt{2}) \sin \tfrac{1}{2}\theta_{ab}\beta_+ + (1/\sqrt{2}) \cos \tfrac{1}{2}\theta_{ab}\beta_-$$
$$\alpha_+ = (1/\sqrt{2}) \cos \tfrac{1}{2}\theta_{ab}\gamma_+ + (1/\sqrt{2}) \sin \tfrac{1}{2}\theta_{ac}\gamma_-$$
$$\alpha_- = -(1/\sqrt{2}) \sin \tfrac{1}{2}\theta_{ac}\gamma_+ + (1/\sqrt{2}) \cos \tfrac{1}{2}\theta_{ac}\gamma_- .$$

Similar expressions hold for the representation of β_+, β_- in terms of α_+, α_- or γ_+, γ_-, and for the representation of γ_+, γ_- in terms of α_+, α_- or β_+, β_-.

This means that

$$p_{\alpha_-}(c = c_+) = \tfrac{1}{2} \sin^2 \tfrac{1}{2}\theta_{ac}$$
$$p_{\alpha_-}(b = b_+) = \tfrac{1}{2} \sin^2 \tfrac{1}{2}\theta_{ab}$$
$$p_{\beta_-}(c = c_+) = \tfrac{1}{2} \sin^2 \tfrac{1}{2}\theta_{bc} .$$

If we take the probability $p_{\alpha_-}(c = c_+)$ as the conditional probability of the value c_+ for the magnitude C, given that the value of A is a_-, then this is computed in the probability space as

$$\frac{p(a = a_- \,\&\, c = c_+)}{p(a_-)}.$$

Similarly

$$p_{\alpha_-}(b = b_+) = \frac{p(a = a_- \,\&\, b = b_+)}{p(a_-)}$$

$$p_{\beta_-}(c = c_+) = \frac{p(b = b_- \,\&\, c = c_+)}{p(b_-)}.$$

From the inequality

$$p(a = a_- \,\&\, c = c_+) \leqslant p(a = a_- \,\&\, b = b_+) + p(b = b_- \,\&\, c = c_+)$$

it follows that

$$\frac{p(a = a_- \,\&\, c = c_+)}{p(a_-)} \leqslant \frac{p(a = a_- \,\&\, b = b_+)}{p(a_-)} + \frac{p(b = b_- \,\&\, c = c_+)}{p(b_-)}$$

if

$$p(b_-) \leqslant p(a_-).$$

Hence

$$p_{\alpha_-}(c = c_+) \leqslant p_{\alpha_-}(b = b_+) + p_{\beta_-}(c = c_+)$$

and again we have the inequality

$$\sin^2 \tfrac{1}{2}\theta_{ac} \leqslant \sin^2 \tfrac{1}{2}\theta_{ab} + \sin^2 \tfrac{1}{2}\theta_{bc}$$

which cannot be satisfied in general. (Notice also that the condition $p(b_-) \leqslant p(a_-)$ can always be satisfied by choosing an appropriate initial measure, and is independent of Bell's inequality.)

Evidently, the probabilities

$$p_{\alpha_-}(c = c_+), \qquad p_{\alpha_-}(b = b_+), \qquad p_{\beta_-}(c = c_+)$$

cannot be computed by the classical rule for conditional probabilities on the probability space. In order to clarify what is involved here, I shall reformulate the derivation. Let Φ_{a_-}, Φ_{b_+}, Φ_{b_-}, Φ_{c_-} be the subsets of the probability space associated with the values

$$a = a_-, \qquad b = b_+, \qquad b = b_-, \qquad c = c_-$$

i.e. the sets of atomic events or ultrafilters for which the magnitudes A,

B, C have these values. Let μ be any measure on the space. Then

$$p(a = a_- \,\&\, c = c_+) = \mu(\Phi_{a_-} \cap \Phi_{c_+})$$
$$p(b = b_- \,\&\, c = c_+) = \mu(\Phi_{b_-} \cap \Phi_{c_+})$$
$$p(a = a_- \,\&\, b = b_+) = \mu(\Phi_{a_-} \cap \Phi_{b_+}).$$

Now, for any subsets Φ_1, Φ_2, Φ_2' on the probability space (where Φ_2' is the set-theoretical complement of Φ_2) and any measure μ:

$$\mu(\Phi_1) = \mu(\Phi_1 \cap \Phi_2) + \mu(\Phi_1 \cap \Phi_2').$$

Hence

$$\mu(\Phi_s \cap \Phi_t) = \mu(\Phi_s \cap \Phi_t \cap \Phi_u) + \mu(\Phi_s \cap \Phi_t \cap \Phi_u')$$
$$\mu(\Phi_s \cap \Phi_u) = \mu(\Phi_s \cap \Phi_t \cap \Phi_u) + \mu(\Phi_s \cap \Phi_t' \cap \Phi_u)$$
$$\mu(\Phi_t \cap \Phi_u') = \mu(\Phi_s \cap \Phi_t \cap \Phi_u') + \mu(\Phi_s' \cap \Phi_t \cap \Phi_u')$$

and so

$$\mu(\Phi_s \cap \Phi_t) \leqslant \mu(\Phi_s \cap \Phi_u) + \mu(\Phi_t \cap \Phi_u').$$

Taking s as a_-, t as c_+, u as b_+, we have

$$\mu(\Phi_{a_-} \cap \Phi_{c_+}) \leqslant \mu(\Phi_{a_-} \cap \Phi_{b_+}) + \mu(\Phi_{c_+} \cap \Phi_{b_+}').$$

Since $\Phi_{b_-} = \Phi_{b_+}'$, i.e. the subset for which $b = b_-$ is the complement of the subset for which $b = b_+$, the inequality becomes

$$\mu(\Phi_{a_-} \cap \Phi_{c_+}) \leqslant \mu(\Phi_{a_-} \cap \Phi_{b_+}) + \mu(\Phi_{b_-} \cap \Phi_{c_+})$$

and so

$$p(a = a_- \,\&\, c = c_+) \leqslant p(b = b_- \,\&\, c = c_+) + p(a = a_- \,\&\, b = b_+).$$

Now this is an inequality between joint probabilities for the values of *incompatible* magnitudes defined on the classical probability space. (Recall that the representation of the statistical states of quantum mechanics by measures on a classical probability space cannot exclude the existence of measures which do not correspond to quantum mechanical statistical states, e.g. dispersion-free measures. The completeness problem concerns the possibility of representing the statistical states as a subset of the full set of measures on a classical probability space.)

It is clear that we cannot compute $p_{a_-}(c = c_+)$ as

$$\frac{\mu(\Phi_{a_-} \cap \Phi_{c_+})}{\mu(\Phi_{a_-})}$$

where μ is the initial probability measure corresponding to the statistical state ψ, say, because

$$p_{a_-}(c=c_+) = p_{\gamma_+}(a=a_-) = \tfrac{1}{2}\sin^2\tfrac{1}{2}\theta_{ac}$$

and

$$\frac{\mu(\Phi_{a_-} \cap \Phi_{c_+})}{\mu(\Phi_{a_-})} \neq \frac{\mu(\Phi_{a_-} \cap \Phi_{c_+})}{\mu(\Phi_{c_+})}$$

unless $\mu(\Phi_{a_-}) = \mu(\Phi_{c_+})$. The crucial assumption in the derivation of Bell's inequality from the inequality for the joint probabilities is that

$$p_{\alpha_-}(c=c_+), \qquad p_{\alpha_-}(b=b_+), \qquad p_{\beta_-}(c=c_+)$$

are to be computed as conditional probabilities

$$\frac{\mu(\Phi_{a_-} \cap \Phi_{c_+})}{\mu(\Phi_{a_-})}, \quad \frac{\mu(\Phi_{a_-} \cap \Phi_{b_+})}{\mu(\Phi_{a_-})}, \quad \frac{\mu(\Phi_{b_-} \cap \Phi_{c_+})}{\mu(\Phi_{b_-})}$$

on the classical probability space in a classical representation of the quantum statistics.

Now, a conditional probability

$$p(s=s' \mid t=t')$$

is determined by a new measure $\mu_{t'}$, defined by

$$\mu_{t'}(\Phi) = \frac{\mu(\Phi \cap \Phi_{t'})}{\mu(\Phi_{t'})}$$

for every measurable set.

The measure $\mu_{t'}$ is the initial measure μ, 'renormalized' to the set $\Phi_{t'}$, i.e. $\mu_{t'}$ satisfies the conditions:

(i) $\qquad \mu_{t'}(\Phi_{t'}) = 1$

(ii) $\qquad$ if $\Phi_u \subseteq \Phi_{t'}$ and $\Phi_v \subseteq \Phi_{t'}$

$\qquad$ then $\dfrac{\mu_{t'}(\Phi_u)}{\mu_{t'}(\Phi_v)} = \dfrac{\mu(\Phi_u)}{\mu(\Phi_v)}.$

Condition (ii) ensures that $\mu_{t'}$ preserves the relative measures of subsets in $\Phi_{t'}$ defined by μ. Notice that

$$\mu_{t'}(\Phi_u) = \frac{\mu(\Phi_u \cap \Phi_{t'})}{\mu(\Phi_{t'})} = \frac{\mu(\Phi_u)}{\mu(\Phi_{t'})}$$

$$\mu_{t'}(\Phi_v) = \frac{\mu(\Phi_v \cap \Phi_{t'})}{\mu(\Phi_{t'})} = \frac{\mu(\Phi_v)}{\mu(\Phi_{t'})}$$

and hence

$$\frac{\mu_{t'}(\Phi_u)}{\mu_{t'}(\Phi_v)} = \frac{\mu(\Phi_u)}{\mu(\Phi_v)}.$$

It follows immediately that if μ is a measure corresponding to some quantum mechanical statistical state, then $\mu_{t'}$ (where $t = t'$ is the value of a quantum mechanical magnitude T) will not in general be a measure corresponding to a quantum mechanical state. Thus,

$$p_{\alpha-}(c = c_+), \qquad p_{\alpha-}(b = b_+), \qquad p_{\beta-}(c = c_+)$$

cannot be computed as conditional probabilities on the classical probability space, and Bell's inequality cannot be derived.

To see that this is so, it suffices to consider pure statistical states. Suppose μ is the initial measure corresponding to the pure statistical state represented by the Hilbert space vector ψ. Let τ' be the eigenvector corresponding to the eigenvalue t' of T, and suppose that τ' does not coincide with ψ, and is not orthogonal to ψ (i.e. τ' and ψ do not form part of any orthogonal set of basis vectors in the Hilbert space, and so do not correspond to the eigenvectors of any common magnitude). Then $\mu_{t'}$ preserves the relative measures of subsets in $\Phi_{t'}$ defined by μ, but $\mu_{\tau'}$ – the measure corresponding to the statistical state τ' – is *uniform* over the set $\Phi_{t'}$. Both measures satisfy condition (i), i.e.

$$\mu_{t'}(\Phi_{t'}) = \mu_{\tau'}(\Phi_{t'}) = 1$$

but they do not define the same relative measures on subsets of $\Phi_{t'}$. (A hidden variable theory with a non-uniform measure $\mu_{\tau'}$ over the set $\Phi_{t'}$ is conceivable. The above argument still applies, for it requires only that $\mu_{t'}$ depends on ψ, while $\mu_{\tau'}$ depends on τ'. Hence $\mu_{\tau'}$ could not in general preserve the relative measures of subsets in $\Phi_{t'}$ as defined by ψ.)

It might be supposed that the situation is different for the composite 2-particle system. A precisely analogous argument applies:

$$p(a^1 = a^1_+ \ \& \ c^2 = c^2_+) = \mu(\Phi_{a+1} \cap \Phi_{c+2})$$
$$p(b^1 = b^1_+ \ \& \ c^2 = c^2_+) = \mu(\Phi_{b+1} \cap \Phi_{c+2})$$
$$p(a^1 = a^1_+ \ \& \ b^2 = b^2_+) = \mu(\Phi_{a+1} \cap \Phi_{b+2}).$$

If Φ_r is a set such that

$$\mu(\Phi_u \cap \Phi_r) = \mu(\Phi'_u \cap \Phi'_r) = 0$$

then

$$\mu(\Phi_t \cap \Phi'_u) = \mu(\Phi_t \cap \Phi_r)$$

for

$$\mu(\Phi_t \cap \Phi'_u) = \mu(\Phi_t \cap \Phi'_u \cap \Phi_r) + \mu(\Phi_t \cap \Phi'_u \cap \Phi'_r)$$
$$\mu(\Phi_t \cap \Phi_r) = \mu(\Phi_t \cap \Phi_r \cap \Phi_u) + \mu(\Phi_t \cap \Phi_r \cap \Phi'_u)$$

and

$$\mu(\Phi_t \cap \Phi'_u \cap \Phi'_r) = \mu(\Phi_t \cap \Phi_r \cap \Phi_u) = 0.$$

Hence the inequality

$$\mu(\Phi_s \cap \Phi_t) \leqslant \mu(\Phi_s \cap \Phi_u) + \mu(\Phi_t \cap \Phi_r)$$

follows from the inequality

$$\mu(\Phi_s \cap \Phi_t) \leqslant \mu(\Phi_s \cap \Phi_u) + \mu(\Phi_t \cap \Phi'_u).$$

Taking s as c^2_+, t as a^1_+, and u as b^1_+, we have

$$\mu(\Phi_{a+1} \cap \Phi_{c+2}) \leqslant \mu(\Phi_{b+1} \cap \Phi_{c+2}) + \mu(\Phi_{a+1} \cap \Phi'_{b+1}).$$

Now, the singlet spin state statistical correlations require that

$$\mu(\Phi_{b+1} \cap \Phi_{b+2}) = \mu(\Phi_{b-1} \cap \Phi_{b-2}) = 0$$

i.e.

$$\mu(\Phi_{b+1} \cap \Phi_{b+2}) = \mu(\Phi'_{b+1} \cap \Phi'_{b+2}) = 0.$$

Hence, taking r as b^2_+, the above inequality becomes

$$\mu(\Phi_{a+1} \cap \Phi_{c+2}) \leqslant \mu(\Phi_{b+1} \cap \Phi_{c+2}) + \mu(\Phi_{a+1} \cap \Phi_{b+2})$$

i.e.

$$p(a^1 = a^1_+ \ \& \ c^2 = c^2_+) \leqslant p(a^1 = a^1_+ \ \& \ b^2 = b^2_+) + p(b^1 = b^1_+ \ \& \ c^2 = c^2_+).$$

At first sight it appears that my objection does not apply to this computation, because the step from the above inequality to Bell's inequality

$$\sin^2 \tfrac{1}{2}\theta_{ac} \leqslant \sin^2 \tfrac{1}{2}\theta_{ab} + \sin^2 \tfrac{1}{2}\theta_{bc}$$

does not depend on any assumptions concerning conditional probabilities. This is an inequality between joint probabilities for the values of *compatible* magnitudes, which is well-defined both on the classical probability space and in Hilbert space. But the 'mirror-image' correlations of the singlet spin state require that

$$p(a^1 = a^1_+ \,\&\, c^2 = c^2_+) = p(a^2 = a^2_-)\, p_{\alpha_{-2}}(c^2 = c^2_+)$$
$$p(b^1 = b^1_+ \,\&\, c^2 = c^2_+) = p(b^2 = b^2_-)\, p_{\beta_{-2}}(c^2 = c^2_+)$$
$$p(a^1 = a^1_+ \,\&\, b^2 = b^2_+) = p(a^2 = a^2_-)\, p_{\alpha_{-2}}(b^2 = b^2_+)$$

with

$$p(a^2 = a^2_\pm) = p(b^2 = b^2_\pm) = p(c^2 = c^2_\pm) = \tfrac{1}{2}$$

and

$$p(a^1 = a^1_\pm) = p(b^1 = b^1_\pm) = p(c^1 = c^1_\pm) = \tfrac{1}{2}.$$

In words: The probability specified by the singlet spin state that A^1 has the value a^1_+ and C^2 has the value c^2_+ is equal to the product of the probability that A^2 has the value a^2_- as defined by the singlet spin state (i.e. $\tfrac{1}{2}$) and the probability, *specified by the statistical state α^2_-*, that C^2 has the value c^2_+. This means that the probabilities

$$p_{\alpha_{-2}}(c^2 = c^2_+), \quad p_{\beta_{-2}}(c^2 = c^2_+), \quad p_{\alpha_{-2}}(b^2 = b^2_+)$$

are to be computed as conditional probabilities:

$$\frac{\mu(\Phi_{a_{+1}} \cap \Phi_{c_{+2}})}{\mu(\Phi_{a_{-2}})}, \quad \frac{\mu(\Phi_{b_{+1}} \cap \Phi_{c_{+2}})}{\mu(\Phi_{b_{-2}})}, \quad \frac{\mu(\Phi_{a_{+1}} \cap \Phi_{b_{+2}})}{\mu(\Phi_{a_{-2}})}$$

where the measure μ corresponds to the singlet spin state.
Since

$$\mu(\Phi_{a_{+2}} \cap \Phi_{a_{+1}}) = \mu(\Phi_{a_{-1}} \cap \Phi_{a_{-2}}) = 0$$

i.e.

$$\mu(\Phi_{a_{+2}} \cap \Phi_{a_{+1}}) = \mu(\Phi'_{a_{+2}} \cap \Phi'_{a_{+1}}) = 0$$

it follows that

$$\mu(\Phi_{a_{+1}} \cap \Phi_{c_{+2}}) = \mu(\Phi_{a_{-2}} \cap \Phi_{c_{+2}}).$$

To see this, recall that for any measure μ, if

$$\mu(\Phi_u \cap \Phi_r) = \mu(\Phi'_u \cap \Phi'_r) = 0$$

then
$$\mu(\Phi_t \cap \Phi_u') = \mu(\Phi_t \cap \Phi_r).$$

Take r as a_+^1, u as a_+^2, and t as c_+^2, and the above relation follows. Similarly:

$$\mu(\Phi_{b_{+1}} \cap \Phi_{c_{+2}}) = \mu(\Phi_{b_{-2}} \cap \Phi_{c_{+2}})$$
$$\mu(\Phi_{a_{+1}} \cap \Phi_{b_{+2}}) = \mu(\Phi_{a_{-2}} \cap \Phi_{b_{+2}}).$$

It follows that the Bell-Wigner computation implicitly assumes that the probabilities:

$$P_{\alpha_{-2}}(c^2 = c_+^2), \quad P_{\beta_{-2}}(c^2 = c_+^2), \quad P_{\alpha_{-2}}(b^2 = b_+^2)$$

are to be computed as the conditional probabilities

$$\frac{\mu(\Phi_{a_{-2}} \cap \Phi_{c_{+2}})}{\mu(\Phi_{a_{-2}})}, \quad \frac{\mu(\Phi_{b_{-2}} \cap \Phi_{c_{+2}})}{\mu(\Phi_{b_{-2}})}, \quad \frac{\mu(\Phi_{a_{-2}} \cap \Phi_{b_{+2}})}{\mu(\Phi_{a_{-2}})}.$$

This assumption is exactly analogous to the objectionable assumption in the case of the single spin-$\frac{1}{2}$ particle.

To sum up: I have shown that the Bell-Wigner argument excludes a classical representation of the quantum statistics on the basis of an obviously untenable assumption concerning the correspondence between quantum statistical states and their representative measures on the probability space. In particular, the argument has nothing whatsoever to do with locality.

On the basis of Bell's argument, Clauser, Horne, Shimony, and Holt designed an experiment to test the validity of Bell's inequality for the spatially separated components of a certain composite system. The experiment was proposed as a crucial test between quantum mechanics and the class of 'local' hidden variable theories. It was actually carried out (by Freedman and Clauser), and the statistical correlations of quantum mechanics were confirmed. Although Holt later obtained results in conflict with the quantum statistics, it is now generally assumed that only 'non-local' hidden variable theories are viable alternatives to quantum mechanics. Clearly, these experiments prove nothing of any theoretical interest at all. If Holt's results are confirmed, and Bell's inequality is found to hold for arbitrary angles, the statistics of a single spin-$\frac{1}{2}$ particle would be unexplained, and no 'local' hidden variable theory could possibly account for the violation of the inequality in this case.

CHAPTER VII

RESOLUTION OF THE COMPLETENESS PROBLEM

The theorem proved by Kochen and Specker is the non-imbeddability of the partial Boolean algebra of idempotent magnitudes – propositions – of quantum mechanics into a Boolean algebra, in the case of systems associated with Hilbert spaces of three or more dimensions. This means the impossibility of representing the statistical states of the quantum algorithm as probability measures on a classical probability space, in such a way that the structure of the set of (idempotent) magnitudes is preserved. Now, the structure involved here is the compatibility structure. One might raise the following objection to the Kochen and Specker result as a completeness proof for the quantum statistics:

Suppose compatibility is defined as in Chapter I, i.e. two magnitudes, A_1 and A_2, are compatible just in case there exists a magnitude, B, and two functions, $g_1 : R \to R$ and $g_2 : R \to R$ such that

$$A_1 = g_1(B)$$
$$A_2 = g_2(B).$$

If, in general, $g(B)$ is understood as the magnitude corresponding to the operator

$$g(B) = \int g(r) \, dP_B(r)$$

this amounts to a definition of $g(B)$ as that magnitude M satisfying the relation

$$p_W(m \in S) = p_W(b \in g^{-1}(s))$$

for every statistical state W of quantum mechanics, and every Borel set S, on the assumption that two magnitudes, A_1 and A_2 are equivalent if and only if:

$$p_W(a_1 \in S) = p_W(a_2 \in S)$$

for every W, S. (Here, g is again a real-valued function on the real line. The symbol a_i is a variable denoting a general value of the magnitude A_i, not a name for the i'th eigenvalue of a magnitude A as in Chapter I, Section IV. Thus, $a_i \in S$ is to be read: The value of the magnitude A_i lies in the range S.)

Thus, two magnitudes, A_1 and A_2, are compatible just in case there exists a magnitude, B, and two functions, $g_1 : R \to R$, $g_2 : R \to R$ such that:

$$p_W(a_1 \in S) = p_W(b \in g_1^{-1}(S))$$
$$p_W(a_2 \in S) = p_W(b \in g_2^{-1}(S))$$

for every W, S.

Now, with this definition of equivalence, functional relationship, and compatibility, surely the theorem of Kochen and Specker is unsatisfactory as a completeness proof for the quantum theory, because part of the problem at issue is just whether the set of statistical states, $\{W\}$, generates *all possible* probability measures, and in what sense, since it is easily conceivable that two *different* magnitudes, A_1 and A_2, represented by *different* Hilbert space operators, satisfy the condition

$$p_W(a_1 \in S) = p_W(a_2 \in S)$$

for every quantum statistical state W and every Borel set S? The magnitudes would then be statistically equivalent with respect to the set $\{W\}$ of statistical states of quantum mechanics, but inequivalent in the algebra of magnitudes represented by self-adjoint Hilbert space operators. And the magnitude M defined by the condition

$$p_W(m \in S) = p_W(b \in g^{-1}(S))$$

for every quantum statistical state W, and every Borel set S, would not necessarily be equivalent to the magnitude $g(A)$ defined by this condition with respect to a wider class of statistical states which are complete in some sense for the algebra of magnitudes.

To put this objection another way: The Kochen and Specker proof shows only that the mere statistical equivalence of two magnitudes, A_1 and A_2, for the statistical ensembles of quantum mechanics does not guarantee equivalence with respect to the values of these magnitudes for individual systems in an ensemble. The reconstruction of the quantum statistics on a classical probability space does not require the statistical equivalence of the random variables representing A_1 and A_2 for *all* probability distributions. In particular, f_{A_1} and f_{A_2} need not be statistically equivalent for dispersion-free distributions, i.e. f_{A_1} and f_{A_2} may take on different values at the same point in probability space. All we require is the statistical equivalence of f_{A_1} and f_{A_2} for those probability measures corresponding to the statistical states of the quantum theory. Only by *assuming* that equivalence in the partial algebra of magnitudes defined relative to the set of quantum states is not merely statistical can we claim the Kochen and Specker theorem as a proof of the impossibility of representing the statistical states of the theory as measures on a classical probability space. We might equally well assume that such a representation is always possible (or, at least, possible in the case of quantum mechanics), in which case it follows that this equivalence in the partial algebra of magnitudes is merely statistical, and cannot be extended to the representative random variables.

That is to say, there are points, x, in the probability space X such that:

$$f_{A_1}(x) \neq f_{A_2}(x)$$

and measures, ϱ, such that

$$\varrho(f_{A_1}^{-1}(s)) \neq \varrho(f_{A_2}^{-1}(s))$$

for some Borel sets S, but

$$\varrho_W(f_{A_1}^{-1}(s)) = \varrho_W(f_{A_2}^{-1}(s))$$

for all measures ϱ_W corresponding to statistical states of the quantum theory.

Now, of course, a set of statistical states assigning probabilities to ranges of values of a set of magnitudes can always be represented as probability measures on a classical probability space, if there are no structural constraints on the representation of the magnitudes as random variables. I discussed the possibility of such a construction at the beginning of Chapter V. It is clear, however, that the formal possibility of this representation is completely uninteresting theoretically. The force of the argument derives from the presumption that the structural constraints imposed here reflect equivalence relationships valid only for a restricted set of statistical states. But this is a misunderstanding. The problem concerns the character of the statistics definable on a given class of event structures, specified by the algebraic structures of the idempotent magnitudes of quantum mechanical systems. And this is given by the partial

Boolean algebras of subspaces or projection operators of the Hilbert spaces of the theory.

Kochen and Specker contribute to this misunderstanding by pointing out in the introductory sections of their article that equivalence, functional relationship, and compatibility in the algebra of magnitudes can be defined with respect to the set of quantum states. If linear combinations and products are defined for compatible magnitudes by

$$c_1 A_1 + c_2 A_2 = (c_1 g_1 + c_2 g_2)(B)$$
$$A_1 A_2 = (g_1 g_2)(B)$$

where $A_1 = g_1(B)$, $A_2 = g_2(B)$, the set of quantum magnitudes acquires the structure of a partial algebra. The condition

$$f_{g(A)} = g(f_A)$$

on the association $A \to f_A$, representing the magnitudes as random variables on a probability space X, preserves the structure of the partial algebra:

$$
\begin{aligned}
f_{c_1 A_1 + c_2 A_2} &= f_{(c_1 g_1 + c_2 g_2) B} \\
&= (c_1 g_1 + c_2 g_2) f_B \\
&= c_1 g_1(f_B) + c_2 g_2(f_B) \\
&= c_1 f_{g_1(B)} + c_2 f_{g_2(B)} \\
&= c_1 f_{A_1} + c_2 f_{A_2}
\end{aligned}
$$

and

$$
\begin{aligned}
f_{A_1 A_2} &= f_{(g_1 g_2)(B)} \\
&= (g_1 g_2)(f_B) \\
&= g_1(f_B) g_2(f_B) \\
&= f_{g_1(B)} f_{g_2(B)} \\
&= f_{A_1} f_{A_2}.
\end{aligned}
$$

What this condition says is that the value assigned to the magnitude $g(A)$ is always equal to the value derived by applying the function g to the value assigned to the magnitude A. Thus, the imbedding of a partial algebra of quantum magnitudes into a commutative algebra of random variables on a probability space might be understood as a map, from the set of magnitudes into a set of random variables, which preserves the functional relationships between magnitudes. And the Kochen and

Specker proof then amounts to a demonstration that no such map exists. Specifically: If the compatible magnitudes A_1, A_2,... are all expressible as functions of a magnitude B, i.e.

$$A_1 = g_1(B), \qquad A_2 = g_2(B), \dots .$$

then if B is represented by the random variable f_B, A_1 is represented by the random variable $f_{g_1(B)} = g_1(f_B)$, A_2 is represented by the random variable $f_{g_2(B)} = g_2(f_B)$, etc. – this cannot be achieved.

Now, this way of presenting the significance of the Kochen and Specker proof is misleading, because the algebraic structure generated by defining equivalence, functional relationship, and compatibility with respect to the set of statistical states of quantum mechanics is isomorphic to the partial algebra of self-adjoint Hilbert space operators *just because* this set of states is *complete* for the partial algebra of operators. And the completeness of the quantum statistics is what is at issue here. The contribution of Kochen and Specker lies in showing that the problem of hidden variables is not that of fitting a theory – i.e. a class of event structures – to a statistics. This can always be done in an infinite number of ways; in particular, a Boolean representation is always possible. Rather, the problem concerns the kind of statistics definable on a given class of event structures. The Kochen and Specker proof is a demonstration that the statistics definable on the event structures of quantum mechanics is not representable by probability measures on a classical probability space. The completeness of the statistics generated by the algorithm of quantum mechanics for this class of structures is shown by Gleason's theorem.

Gleason's theorem was proposed as a solution to a problem posed by Mackey: to specify all possible measures on the subspaces of a Hilbert space, where a measure is a map, μ, from the subspaces onto the non-negative real numbers satisfying the additivity condition

$$\mu\left(\bigvee_i \mathscr{H}_i\right) = \sum_i \mu(\mathscr{H}_i)$$

for a countable set of mutually orthogonal subspaces. (Here $\bigvee_i \mathscr{H}_i$ is the span of the subspaces $\{\mathscr{H}_i\}$, the subspace which is the least upper bound of the set.)

The theorem states that in a Hilbert space of three or more dimensions, every measure is representable as

$$\mu\,(\mathscr{K}) = \mathrm{Tr}\,(WP)$$

where W is a statistical operator, and P is the projection operator whose range is $\mathscr{K}$.

If we define a generalized probability measure as a normed, countably additive, real-valued function, μ, on a *partial Boolean algebra*, i.e. a measure function satisfying the usual conditions for a probability measure on each maximal compatible subset of the partial Boolean algebra, then the probability assignments generated by the algorithm of quantum mechanics are generalized probabilities in this sense, and Gleason's theorem shows that the quantum algorithm generates all possible generalized probability measures on the partial Boolean algebra of propositions of a quantum mechanical system.

From a purely formal point of view the Kochen and Specker result follows as a corollary to Gleason's theorem. A dispersion-free probability measure in the generalized sense is a 2-valued homomorphism, and so the impossibility of imbedding a partial Boolean algebra of quantum propositions into a Boolean algebra follows from the non-existence of dispersion-free states. However, the significance of Gleason's theorem for the completeness problem of quantum mechanics is only fully brought out by Kochen and Specker's notion of a partial Boolean algebra, which completely clarifies the sense in which the Boolean event structures of classical mechanics are generalized by quantum mechanics.

The core of Bell's objection to the completeness proofs of von Neumann, Jauch and Piron, and Kochen and Specker (in the form of a corollary to Gleason's theorem) is essentially the argument I have discussed above. (Bell (a), p. 447).

It will be urged that these analyses leave the real question untouched. In fact it will be seen that these demonstrations require from the hypothetical dispersion free states, not only that appropriate ensembles thereof should have all measurable properties of quantum mechanical states, but certain other properties as well. These additional demands appear reasonable when results of measurement are loosely identified with properties of isolated systems. They are seen to be quite unreasonable when one remembers with Bohr "the impossibility of any sharp distinction between the behaviour of atomic objects and the interaction with the measuring instruments which serve to define the conditions under which the phenomena appear."

The reference to Bohr here serves to legitimize a dispositional interpretation of the quantum mechanical magnitudes. The 'result of measurement', i.e. the assignment of a value to an idempotent magnitude, is only 'loosely identified' with a property of the system, for it represents the disposition of the system to function in a certain way under certain conditions defined by a macroscopic measuring instrument. Since incompatible magnitudes A, B represent dispositions referring to incompatible conditions, the equality

$$\mathrm{Exp}_W (A + B) = \mathrm{Exp}_W (A) + \mathrm{Exp}_W (B)$$

for the set of statistical states of quantum mechanics should be regarded as a relation peculiar to the theory. It would therefore be unreasonable to require that this condition be satisfied by dispersion-free states in the representation of the statistical states of the theory by measures on a classical probability space. This is Bell's objection to von Neumann's proof. His objections to the Jauch and Piron proof and the Kochen and Specker proof are similar. He shows that these proofs implicitly impose restrictions on the values assigned to incompatible magnitudes by dispersion-free measures on the classical probability space, and that these restrictions are unreasonable on a Bohrian dispositional interpretation of the magnitudes.

In the case of the Kochen and Specker proof (or Gleason's theorem, for Bell), the intransitivity of the compatibility relation is exploited. Bell sees the relevant corollary to Gleason's theorem as providing a proof of von Neumann's result without the additivity condition for the expectation values of incompatible magnitudes. Since compatibility is intransitive, even the additivity condition for compatible magnitudes implicitly imposes restrictions on incompatible magnitudes.

For example, consider the magnitudes A and C represented by the operators:

$$A = a_1 A_1 + a_2 A_2 + a_3 A_3$$
$$C = c_1 C_1 + c_2 C_2 + c_3 C_3$$

where

$$A_1 = C_1 = P$$
$$A_2 = Q, \qquad A_3 = R$$
$$C_2 = Q', \qquad C_3 = R'$$

and P, Q, R are projection operators onto 3 mutually orthogonal 1-dimensional subspaces in $\mathscr{H}_3$, and Q', R' are projection operators onto a different orthogonal pair of subspaces in the plane defined by Q and R. Both A and C are compatible with the magnitude

$$B = b_1 B_1 + b_2 B_2$$

where

$$B_1 = P$$
$$B_2 = P^{\perp}$$

but A and C are incompatible.

For Bell, the idempotent magnitudes A_1 and C_1 represent dispositions referring to the behaviour of the system under conditions defined by an A-instrument and a C-instrument, respectively, which are mutually incompatible. Although

$$p_W(a = a_1) = p_W(b = b_1) = p_W(c = c_1)$$

for all statistical states W of quantum mechanics, this equality cannot be required for the dispersion-free states, for this would amount to stipulating that A_1 and C_1 represent the same dispositions. A_1 and C_1 are statistically equivalent for the statistical states of quantum mechanics, and are represented by the same projection operator, P, in the theory. Bell suggests that in a hidden variable theory, A_1 and C_1 should be represented by *different* random variables on the probability space, which are statistically equivalent only for those measures corresponding to the statistical states of quantum mechanics.

The dispositional interpretation of the magnitudes serves only to motivate Bell's proposal that equivalence in the algebra of quantum magnitudes need not be preserved in a hidden variable theory. I have argued above that the question of the possibility of a Boolean representation of the quantum statistics is uninteresting if the algebraic structure of the magnitudes of the theory is not preserved, e.g. if the equivalence of A_1 and C_1 is not preserved. The completeness problem makes sense only with respect to a given class of structures.

CHAPTER VIII

THE LOGIC OF EVENTS

I. REMARKS

In the previous two chapters, I have shown that the statistical relations of quantum mechanics cannot be represented by measures on a classical probability space, because the partial algebra of quantum magnitudes is not imbeddable in a commutative algebra. Thus, the quantum statistics cannot be interpreted as an incomplete classical (i.e. Boolean) statistics. Moreover, the set of statistical states of the theory generates all possible probability measures in the generalized sense on the partial Boolean algebra of idempotent magnitudes. The purpose of this chapter, and the next, is to clarify the significance of non-imbeddability here. My thesis is that the transition from classical to quantum mechanics involves the generalization of the Boolean event structures of classical mechanics to non-Boolean event structures of a particular kind.

In order to develop this thesis, I need certain elementary notions of classical logic. In Section II of this chapter, I shall give an exposition of the proof theory (syntax) and model theory (semantics) of classical first-order logic, the predicate calculus. First-order logic concerns the formalization of propositions which deal with the quantification of individuals in a certain domain, with respect to properties and relations predicated of these individuals. (Higher-order logics involve quantification over properties and relations.) I discuss the proof theory of this logic, the notion of theoremhood, or provability from a set of axioms, and the semantic notion of truth under an interpretation for the sentences of the formal system. Classical first order logic is complete: the proof theory generates as theorems all (and only) the universally valid sentences, i.e. those sentences which are true in every interpretation.

I prove this theorem for the classical propositional calculus as an application of Stone's representation theorem for Boolean algebras: every Boolean algebra is isomorphic to a perfect, reduced field of sets. The propositional calculus treats the logical operations of negation ('not'), conjunction ('and'), disjunction ('or'), etc. as truth functions of the atomic propositions. Syntactically, the atomic propositions may be regarded as 0-place predicates, which are assigned truth values under an interpretation independently of the relations holding between individuals in the set-theoretic interpretation of the predicate calculus. The completeness proof requires the algebraic formulation of the propositional calculus, via the notion of the Lindenbaum-Tarski algebra of a logic. The basic idea of Stone's theorem involves the concept of an ultrafilter, essentially a maximal consistent set of propositions, which is picked out be a possible assignment of truth values to the atomic propositions. An ultrafilter, then, represents a set of events which is maximal in a logical sense: all events belong to the set which are consistent with the existence of a particular set of atomic events and the non-existence of all other atomic events. The Lindenbaum-Tarski algebra of the classical propositional calculus is a Boolean algebra. The relevant corollary to Stone's theorem says, in effect, that every *possible* event is represented by a proposition that belongs to *at least one ultrafilter*. The isomorphism associates events or propositions with the set of ultrafilters containing the proposition.

In Section III, I shall argue that the phase space of classical mechanics is a topological representation of a Boolean event structure, while the Hilbert space of quantum mechanics represents a strongly non-Boolean event structure.

II. CLASSICAL LOGIC

The formalization of any mathematical domain – e.g. arithmetic – in terms of classical logic involves a formal language, L, with variables (x, y, z, etc., ranging over individuals in the domain in the case of a first-order theory), predicates or relational symbols (F, G, H, etc., representing properties and relations), connectives ($\land$, $\neg$, etc., for conjunction, negation, etc.), and quantifiers (the existential quantifier $\exists x$ – read 'there exists an x' – and the universal quantifier $\forall x$ – read 'for all x'). Out of these symbols formulae are constructed. For example, $Fx \land Gy$ is a well-formed formula in L. Read this as: x has the property F and y has the property G. The formula $\neg Hxy$ is read as: it is not the case that x and y stand in the relation H. Quantifying over x and y, we form formulae such as: $(\exists x)\,(\forall y)\,(Fx \wedge Gy)$ – there is an x such that for every y, x has the property F and y has the property G.

The well-formed formulae are generated from the atomic formulae by the logical operations of negation, conjunction, and existential quantification. An *atomic formula* is of the form $Fx_1...x_n$, where F is a relational symbol of degree n. Other connectives may be defined in terms of negation and conjunction. The disjunction, $s \vee t$, where s and t are any formulae, is defined as $\neg(\neg s \wedge \neg t)$; the conditional, $s \supset t$, is defined as $\neg(s \wedge \neg t)$, i.e. $\neg s \vee t$; the biconditional, $s \equiv t$, is defined as $(s \supset t) \wedge (t \supset s)$. The universal quantifier may be defined in terms of the existential quantifier and negation: $(\forall x)s$ is $\neg(\exists x)\neg s$. An occurrence of a variable in a formula is said to be *bound* if it is governed by a quantifier; otherwise it is free. For example, in the formula $(\forall x)Fxy \wedge Gx$, the occurrence of the variable x in the subformula Fxy is bound, while the occurrence of x in the subformula Gx is free. The occurrence of y is free. In the formula $(\exists x)\,(Fxy \wedge Gx)$, both occurrences of x are bound. A formula in which no variable occurs free is a *sentence*.

A formal theory – say, of arithmetic – requires a finite number of *axioms* (certain selected formulae in L) and a finite number of *inference rules*, which allow *theorems* to be generated from the axioms in a *proof*.

An *interpretation* of a formal theory is a *relational structure*:

$$\mathcal{M} = (M\,; R, S, ...)$$

where M is a non-empty set, and $R, S, ...$ are properties and relations on M. Consider an assignment of members of M to the variables of L. A formula $s(x, y, ...)$ with the free variables $x, y, ...$ is said to be *satisfied in* $\mathcal{M}$ by the assignment of m_1 to x, m_2 to y, etc., if the relation over M corresponding to s holds between the elements m_1, m_2, etc. when each relational symbol $F, G, ...$ is replaced by an appropriate relation $R, S, ...$. Notation:

$$\mathcal{M} \vDash s\,[m_1, m_2, ...]$$

where the sequence $m_1, m_2, ...$ is an assignment of elements in M to the free variables in s. For example, consider the formula Fxy and the interpretation $\mathcal{M} = (N; <)$, where N is the set of natural numbers and $<$ is the relation 'less than'. If F is interpreted as the relation $<$, the assignment $[1, 2]$ satisfies Fxy, while the assignment $[4, 3]$ fails to satisfy Fxy.

The notion of satisfaction of a formula in a relational structure $\mathcal{M} = (M; R, S, ...)$ may be defined recursively as follows:

(i) $\mathcal{M} \vDash Fx_1 \cdots x_n[a_1, ..., a_n]$ if and only if the relation R assigned to F holds between the elements $a_1, ..., a_n$, i.e. if and only if the sequence $a_1, ..., a_n$ is in the relation R assigned to F under the interpretation. This defines the notion of satisfaction for atomic formulae in L.

(ii) $\mathcal{M} \vDash \neg s[a_1, a_2, ...]$ if and only if it is not the case that $\mathcal{M} \vDash s[a_1, a_2, ...]$.

(iii) $\mathcal{M} \vDash (s \wedge t)[a_1, a_2, ...]$ if and only if $\mathcal{M} \vDash s[a_1, a_2, ...]$ and $\mathcal{M} \vDash t[a_1, a_2, ...]$.

(iv) $\mathcal{M} \vDash (\exists y)s(x_1, ..., x_n, y)[a_1, ..., a_n]$ if and only if there is a $b \in M$ such that $\mathcal{M} \vDash s(x_1, ..., x_n, y)[a_1, ..., a_n, b]$.

If $\mathcal{M} \vDash s[a_1, ..., a_n]$ for *all* assignments of elements in M to the free variables in s, then s is said to be *true* in $\mathcal{M}$, and $\mathcal{M}$ is said to be a *model* of s. Notation: $\mathcal{M} \vDash s$. The formula s is said to be *universally valid* or *logically true* if $\mathcal{M} \vDash s$ for every interpretation $\mathcal{M}$. Notation: $\vDash s$. For example, the formula $(\exists x)Fxy$ is not true under the interpretation $\mathcal{M} = (N; <)$; the relational structure is not a model for this formula. The relational structure $\mathcal{M} = (I; <)$, where I is the set of positive and negative integers, is a model for $(\exists x)Fxy$. The sentence $(\exists x)(\exists y)(Fxy \supset Fxy)$ is true in every relational structure, i.e. every sequence a, b satisfies the formula $Fxy \supset Fxy$ in every domain, whatever relation is associated with F.

Notice that if s is a sentence, a formula in which no variables occur free, then $\mathcal{M} \vDash s$ if and only if $\mathcal{M} \vDash s[a_1, ..., a_n]$ for some sequence of elements $a_1, ..., a_n$ in M. (In this case, either no sequences satisfy s – vacuously – or all sequences satisfy s.) $\mathcal{M}$ is said to be a model of a set of sentences Σ if and only if $\mathcal{M} \vDash s$ for every $s \in \Sigma$.

A set of axioms sufficient to generate all and only the universally valid formula in L via one or more inference rules is a *proof theory* for the classical predicate calculus. A proof is a finite sequence of formulae, such that each formula is either an axiom or follows from formulae earlier in the sequence by the inference rules. Notation: $\vdash s$ signifies s is provable, or s is a theorem, i.e. s is the last formula in a proof sequence. A proof theory is *sound* if every theorem is a universally valid formula, i.e. if the axioms and inference rules do not generate any formulae which are outside the class of universally valid formulae. A proof theory is *complete* if every universally valid formula is a theorem, i.e. if the axioms and inference rules generate all universally valid formulae. Soundness and completeness are metalogical notions relating the proof theory or syntax of a formal system with the model theory or semantics of the system.

Soundness – if $\vdash s$ then $\vDash s$ – is usually relatively easy to prove. All that needs to be shown is that each of the axioms is universally valid, and that the inference rules preserve universal validity. From this it follows that the set of axioms is consistent, i.e. s and $\neg s$ cannot both be proved in the system, for any s. For if $\vdash s$, then $\vDash s$; and if $\vdash \neg s$, then $\vDash \neg s$. But s and $\neg\ s$ cannot both be universally valid.

In order to prove completeness, it is necessary to show that if $\vDash s$, then $\vdash s$. I shall prove completeness for the classical *propositional* calculus. The propositional calculus may be regarded as a restriction of the predicate calculus to 0-place predicates. In other words, the universally valid formulae of the propositional calculus – the *tautologies* – correspond to those sentences of the predicate calculus whose validity is independent of quantification.

The formulae of the propositional calculus may be generated from a countable set of atomic sentences (0-place predicates) $\{P, Q, R, ...\}$ by negation and conjunction, the disjunction, conditional and biconditional being defined in terms of these connectives.

An appropriate set of axioms is the set of all sentences of any of the forms (Bell and Slomsen, pp. 36, 37):

(1)	$s \supset (t \supset s)$
(2)	$(s \wedge t) \supset s \qquad (s \wedge t) \supset t$
(3)	$s \supset (s \vee t) \qquad t \supset (s \vee t)$
(4)	$(-s \supset -t) \supset (t \supset s)$
(5)	$(s \supset (t \supset u)) \supset ((s \supset t) \supset (t \supset u))$
(6)	$(s \supset t) \supset ((s \supset u) \supset (s \supset (t \wedge u)))$
(7)	$(s \supset t) \supset ((u \supset t) \supset ((s \vee u) \supset t))$.

Theorems are generated from these axioms by the inference rule known as modus ponens: $\{s, s \supset t\} \vdash t$ (t is provable from the set of formulae $\{s, s \supset t\}$).

I have previously introduced the symbols $\wedge$ and $\vee$ for infimum and supremum in a lattice, as well as for meet and join in a Boolean algebra. Considered as a partially ordered set, a Boolean algebra is a complemented, distributive lattice, with the Boolean meet and join corresponding to the lattice infimum and supremum. And the classical propositional calculus is Boolean, in the sense that the Lindenbaum-Tarski algebra of the logic is a Boolean algebra. (Of course, it will always be obvious from the context whether the symbols $\wedge$ and $\vee$ denote lattice operations or logical connectives.)

The Lindenbaum-Tarski algebra $\mathscr{L}$ of a logic is generated by first defining an equivalence relation $\sim$ on the formulae in L:

$$s \sim t \quad \text{if and only if} \quad \vdash s \supset t \quad \text{and} \quad \vdash t \supset s$$

and then defining a relation (transitive, reflexive, and antisymmetric) on the set of equivalence classes of formulae $\{|s| : s \in \mathscr{L}\}$, where $|s| = \{t \in \mathscr{L} : s \sim t\}$:

$$|s| \leqslant |t| \quad \text{if and only if} \quad \vdash s \supset t.$$

The partially ordered set of equivalence classes of formulae is the Lindenbaum-Tarski algebra of the logic.

In the case of the classical propositional calculus, it is easy to verify that the relation $\leqslant$, defined by

$$|s| \leqslant |t| \quad \text{if and only if} \quad \vdash s \supset t$$

is transitive, reflexive, and antisymmetric, i.e. a partial ordering. Now, the infimum of $|s|$ and $|t|$ always exists and is equal to $|s \wedge t|$, because

$$\vdash (s \wedge t) \supset s$$

and

$$\vdash (s \wedge t) \supset t$$

by axiom 2, so that

$$|s \wedge t| \leqslant |s|$$

and

$$|s \wedge t| \leqslant |t|$$

i.e. $|s \wedge t|$ is a lower bound of $|s|$ and $|t|$.

But if, for some u, $|u| \leqslant |s|$, $|u| \leqslant |t|$, then $\vdash u \supset s$ and $\vdash u \supset t$, so that

$$\vdash u \supset (s \wedge t)$$

by axiom 6 and modus ponens, i.e.

$$|u| \leqslant |s \wedge t|.$$

Hence, the greatest lower bound of $|s|$ and $|t|$ is $|s \wedge t|$, i.e.

$$|s| \wedge |t| = |s \wedge t|$$

Similarly, the supremum of $|s|$ and $|t|$ is $|s \vee t|$, i.e.

$$|s| \vee |t| = |s \vee t|.$$

So the partially ordered set of equivalence classes of formulae is a lattice.

It follows easily that the distributive law holds, since

$$\vdash s \wedge (t \vee u) \supset (s \wedge t) \vee (s \wedge u)$$

and

$$\vdash (s \wedge t) \vee (s \wedge u) \supset s \wedge (t \vee u)$$

and hence

$$|s \wedge (t \vee u)| = |(s \wedge t) \vee (s \wedge u)|$$

i.e.

$$|s| \wedge (|t| \vee |u|) = (|s| \wedge |t|) \vee (|s| \wedge |u|).$$

The lattice is complemented because $|s| = 1$ if and only if $\vdash s$, and $|s| = 0$ if and only if $\vdash \neg s$. (If $\vdash s$, then, by axiom 1, $\vdash t \supset s$ for every t, and so $|t| \leqslant |s|$ for every t, i.e. $|s| = 1$. And if $|s| = 1$, then for any t, $|t| \leqslant |s|$, and so $\vdash t \supset s$ for every t. Choosing t an axiom or a theorem, it follows that $\vdash s$.)

Since, for every s

$$\vdash s \vee \neg s$$

and

$$\vdash \neg (s \wedge \neg s)$$

it follows that

$$|s| \vee |\neg s| = 1$$

and

$$|s| \wedge |\neg s| = 0$$

so that a complement

$$|s|' = |\neg s|$$

exists for every $|s|$. The Lindenbaum-Tarski algebra of the classical propositional calculus is therefore a complemented distributive lattice, i.e. a Boolean algebra.

The semantic notion of satisfaction for the propositional calculus evidently does not require any explicit consideration of sequences of elements in the relational structures over which the sentences are interpreted. The satisfaction of s in $\mathcal{M}$ is independent of the relations holding between elements in M, and depends only on the truth values assigned to the atomic sentences (0-place predicates) in $\mathcal{M}$. That is to say, the sentences of the propositional calculus are truth functions of the atomic sentences: the truth or falsity of a sentence constructed by the logical operations of negation and conjunction from the atomic sentences is completely determined by the truth values assigned to the atomic sentences. An interpretation amounts to a particular assignment of the truth values *true* and *false* to the atomic sentences, which is then extended recursively to all sentences. Equivalently, an interpretation is a map, v, from the set of atomic sentences onto the zero and unit elements of the 2-element Boolean algebra $\mathcal{L}_2$, with 0 corresponding to *false* and 1 corresponding to *true*, extended recursively to all sentences as follows:

$$v(\neg s) = v(s)'$$
$$v(s \wedge t) = v(s) \wedge v(t)$$
$$v(s \vee t) = v(s) \vee v(t).$$

It is easy to check that these condition on v ensure that $\neg s$ is true if and only if s is false, that a conjunction is true if and only if both conjuncts are true, and that a disjunction is true if and only if at least one disjunct is true.

Notice that the condition for the disjunction is superfluous, and follows from the conditions for negation and conjunction. The symbols $\wedge$ and $\vee$ in $v(s \wedge t), v(s \vee t)$ denote conjunction and disjunction in the logic L. In $v(s) \wedge v(t), v(s) \vee v(t)$ they denote the meet and join in $\mathcal{L}_2$.

Now, the theorems of the classical propositional calculus belong to the equivalence class $|s| = 1$. A sentence s is inconsistent if and only if a sentence t and its negation are both provable from s, i.e. if and only if

$$s \vdash t \wedge \neg t$$

or

$$\vdash s \supset (t \wedge \neg t).$$

Clearly,

$$\vdash (t \wedge \neg t) \supset s$$

for any sentence t, so s is inconsistent if and only if

$$|s| = |t \wedge \neg t|.$$

But

$$|t \wedge \neg t| = |t| \wedge |\neg t| = 0$$

for any t, and so s is inconsistent if and only if $|s|=0$. A sentence s is consistent if and only if $|s| \neq 0$, i.e. if and only if $|\neg s| \neq 1$, or $\neg s$ is not provable.

To show that s is a theorem if and only if s is a tautology, is to show that $|s| = 1$ if and only if $v(s) = 1$ for every map v from the atomic sentences onto $\mathcal{Z}_2$, the 2-element Boolean algebra. That is to say, $|\neg s| = 0$ if and only if $v(\neg s) = 0$ for every map v, for $v(\neg s) = 0$ if and only if $v(s) = 1$. The negation is superfluous here, since the statement holds for every sentence s of the propositional calculus. Thus: $|s| = 0$ if and only if $v(s) = 0$ for every map v, or equivalently, $|s| \neq 0$ if and only if $v(s) = 1$ for some map v from the atomic sentences onto $\mathcal{Z}_2$. The completeness theorem for the classical propositional calculus may therefore be reformulated: A sentence is consistent if and only if it is satisfiable.

I shall prove completeness as the logical analogue of the fundamental representation theorem for Boolean algebras, Stone's theorem, that every Boolean algebra is isomorphic to a perfect reduced field of sets. The proof of this theorem involves the concept of an unltrafilter, which is of importance in the subsequent discussion.

I use the symbol $\mathcal{B}$ for a Boolean algebra. To distinguish the elements of $\mathcal{B}$ from the members of the representative set X, I shall denote elements of $\mathcal{B}$ by the letters $a, b, c, \ldots$ (from the beginning of the alphabet) and elements of X by the letters $x, y, z, \ldots$ (from the end of the alphabet).

A filter in a Boolean algebra is a non-empty subset Φ of $\mathcal{B}$, satisfying the conditions

(i)   if $a, b \in \Phi$, then $a \wedge b \in \Phi$

(ii)   if $a \in \Phi$ and $a \leqslant b$,   then $b \in \Phi$.

Equivalently, a filter may be defined as a non-empty subset Φ of $\mathscr{B}$ such that $a \wedge b \in \Phi$ if and only if $a \in \Phi$ and $b \in \Phi$, or a non-empty subset $\Phi \subseteq \mathscr{B}$ with $1 \in \Phi$ such that $a \in \Phi$, $a' \vee b$ implies $b \in \Phi$. (The dual notion is that of an *ideal*, i.e. a non-empty subset of $\mathscr{B}$ such that the join of any two elements is a member of the set, and every member of $\mathscr{B}$ below any member of the set belongs to the set.)

A *proper* filter is a proper subset of $\mathscr{B}$, i.e. $\Phi \neq \mathscr{B}$, or $0 \in \Phi$. The *principal* filter *generated* by the element $a \in \mathscr{B}$ is the set

$$\{b \in \mathscr{B} : a \leqslant b\}.$$

Consider the set of all filters in $\mathscr{B}$. This set is partially ordered with respect to the relation of set inclusion. An *ultrafilter* (or maximal filter) is a proper filter that is maximal with respect to this ordering i.e. it is not a proper subset of a proper filter in $\mathscr{B}$. It can be shown that the necessary and sufficient condition for a proper filter Φ in $\mathscr{B}$ to be an ultrafilter is that for every $a \in \mathscr{B}$, either $a \in \Phi$ or $a' \in \Phi$ but not both.

In the case of a power set Boolean algebra, i.e. the Boolean algebra of subsets of a set X, the set of subsets containing a specific point $x \in X$ is an ultrafilter in the power set Boolean algebra – the principal ultrafilter generated by the point x.

A map $h : \mathscr{B}_1 \to \mathscr{B}_2$ from $\mathscr{B}_1$ into $\mathscr{B}_2$ is a *homomorphism* if it preserves the algebraic operations, i.e. if for all $a, b, \in \mathscr{B}_1$:

$$h(a \wedge b) = h(a) \wedge h(b)$$
$$h(a \vee b) = h(a) \vee h(b)$$
$$h(a') = h(a)'.$$

It follows that h maps the zero and unit elements of $\mathscr{B}_1$ onto the zero and unit elements of $\mathscr{B}_2$, and that if $a \leqslant b$ in $\mathscr{B}_1$, then $h(a) \leqslant h(b)$ in $\mathscr{B}_2$.

$\mathscr{B}_1$ and $\mathscr{B}_2$ are *isomorphic* if h is one-one and onto, i.e. if $h(\mathscr{B}_1) = \mathscr{B}_2$, where $h(\mathscr{B}_1)$ is the image of $\mathscr{B}_1$ under the map, i.e. the set of elements in $\mathscr{B}_2$ onto which some element of $\mathscr{B}_1$ is mapped by h. A necessary and sufficient condition for a homomorphism to be an isomorphism is that $h(a) = 0$ implies $a = 0$, or $h^{-1}(0)$ contains only the zero of $\mathscr{B}_1$. (If this condition is satisfied and $h(a) = h(b)$, then

$$h(a \wedge b') = h(a) \wedge h(b)' = 0$$

and

$$h(b \wedge a') = h(b) \wedge h(a)' = 0$$

so that

$$a \wedge b' = b \wedge a' = 0$$

and hence $a \leqslant b$ and $b \leqslant a$, i.e. $a = b$. The converse is trivial.) By duality, a homomorphism h is an isomorphism if and only if $h^{-1}(1)$ contains only the unit element of $\mathscr{B}_1$.

If $h:\mathscr{B}_1 \to \mathscr{B}_2$ is a homomorphism, then the set of elements in $\mathscr{B}_1$ mapped onto 1 by h is a filter, i.e. the set $\Phi = \{a \in \mathscr{B}_1 : h(a) = 1\}$ is a filter.

A 2-*valued homomorphism on $\mathscr{B}$* is a homomorphism from $\mathscr{B}$ onto the 2-element Boolean algebra, $\mathscr{L}_2$. The set of elements in $\mathscr{B}$ mapped onto 1 by a 2-valued homomorphism, i.e. $\{a \in \mathscr{B} : h(a) = 1\}$, is an ultrafilter. Conversely, if Φ is an ultrafilter in $\mathscr{B}$, the map $h:\mathscr{B} \to \mathscr{L}_2$, such that $h(a) = 1$ if $a \in \Phi$ and $h(a) = 0$ if $a \in \Phi$, is a 2-valued homomorphism. Hence, there is a one-one correspondence between the set of ultrafilters on $\mathscr{B}$ and the set of 2-valued homomorphisms on $\mathscr{B}$.

A 2-valued homomorphism, $h:\mathscr{L} \to \mathscr{L}_2$, on the Lindenbaum-Tarski algebra of the classical propositional calculus defines an interpretation or model for the propositional calculus, i.e. a map from the atomic sentences onto $\mathscr{L}_2$:

$$v(P) = h(|P|), \qquad v(Q) = h(|Q|), \ldots$$

that may be extended to the set of all sentences

$$v(s) = h(|s|)$$

because h is a homomorphism (and so if $v(s) = h(|s|)$ and $v(t) = h(|t|)$, for s and t, then $v(s \wedge t) = v(s) \wedge v(t)$ and $v(\neg s) = v(s)'$). Conversely, every interpretation corresponds to a 2-valued homomorphism.

It is a theorem that for every proper filter Φ in a Boolean algebra there exists an ultrafilter that includes Φ. For, consider the set of all filters in $\mathscr{B}$ that include (i.e. are extensions of) a particular filter Φ. This set, $\mathscr{F}$, may be partially ordered by set inclusion. Suppose $\{\Phi_i : i \in I\}$ is a chain of filters with respect to this ordering. Then $\bigcup = \bigcup_{i \in I} \Phi_i$ is a filter, an upper bound for the chain. (If $a, b \in \bigcup$, then $a \in \Phi_i$, $b \in \Phi_j$ for some

Φ_i, Φ_j. But either $\Phi_i \subseteq \Phi_j$, or $\Phi_j \subseteq \Phi_i$, and so either $a, b \in \Phi_i$ or $a, b \in \Phi_j$, and hence either $a \wedge b \in \Phi_i$ or $a \wedge b \in \Phi_j$; in either case $a \wedge b \in \bigcup$. If $a \in \Phi$, then $a \in \Phi_i$, for some $i \in I$, and so $b \in \Phi_i \subseteq \bigcup$ for all $b \in \mathscr{B}$ such that $a \leqslant b$. Obviously $0 \notin \bigcup$, because $0 \notin \Phi_i$, for any $i \in I$.) Since $\Phi \subseteq \bigcup$, U belongs to the set of all filters in $\mathscr{B}$ that are extensions of Φ, and so each chain in the partially ordered set has an upper bound in the set. It follows from Zorn's Lemma that there is a maximal element in the set, an ultrafilter that includes Φ.

It is an immediate consequence of this theorem that each non-zero element $a \in \mathscr{B}$ is contained in some ultrafilter (i.e. the ultrafilter that is the maximal extension of the principal filter generated by a), and that if a and b are distinct elements of $\mathscr{B}$, there is an ultrafilter containing a but not b ($a \neq b$ implies that either not $a \leqslant b$ or not $b \leqslant a$, i.e. either $a \wedge b' \neq 0$, or $a' \wedge b \neq 0$, and so either there is an ultrafilter containing a and b' or there is an ultrafilter containing a' and b).

A *field* of sets, $\mathscr{F}$, is a non-empty set of subsets of a fixed set, X, closed with respect to finite unions, intersections, and complements. (This definition is redundant: If $\mathscr{F}$ is closed with respect to finite intersections and complements, it is closed with respect to finite unions; and if $\mathscr{F}$ is closed with respect to finite unions and complements, it is closed with respect to finite meets.) Obviously, a field of sets is a Boolean algebra.

A field $\mathscr{F}$ of subsets of a set X is said to be *reduced* if for every distinct pair of points x, y in X there exists a set in $\mathscr{F}$ containing x but not containing y. $\mathscr{F}$ is *perfect* if every ultrafilter in $\mathscr{F}$ is determined by a point in X. The sets in $\mathscr{F}$ containing a specific point $x \in X$ form an ultrafilter (if a set does not contain x, its complement does), the ultrafilter *determined* by x.

If $\mathscr{F}$ is a perfect reduced field of subsets of a set X, then the one-one correspondence between ultrafilters and 2-valued homomorphisms can be extended to points of X. Every point in X determines a unique ultrafilter (or 2-valued homomorphism), and conversely every ultrafilter (or 2-valued homomorphism) is determined by a point in X. (If x and y are distinct points in X, they determine different ultrafilters, because $\mathscr{F}$ is a reduced field of sets, and so there is a set in $\mathscr{F}$ containing x, i.e. belonging to the ultrafilter determined by x, but not containing y, i.e. not belonging to the ultrafilter determined by y.)

Stone's theorem can now be formulated as follows: If X is a set of ultrafilters in a Boolean algebra $\mathscr{B}$, and $h(a)$ is the subset of ultrafilters in X containing the element $a \in \mathscr{B}$, then the set of all such subsets

$$\mathscr{F} = \{h(a) : a \in \mathscr{B}\}$$

is a perfect reduced field of subsets of X, and h is an isomorphism from $\mathscr{B}$ onto $\mathscr{F}$.

The map $h : \mathscr{B} \to \mathscr{F}$ is a homomorphism, because

$$h(a \wedge b) = h(a) \cap h(b)$$

i.e. the subset of ultrafilters (in X) containing the element $a \wedge b \in \mathscr{B}$ is the set-theoretical intersection of the subsets $h(a)$ and $h(b)$, since $a \wedge b$ belongs to the ultrafilter $\Phi \in X$ if and only if both a and b belong to Φ; and

$$h(a') = h(a')$$

i.e. the subset of ultrafilters containing the element $a' \in \mathscr{B}$ is identical to the set-theoretical complement in X of the subset of ultrafilters containing the element $a \in \mathscr{B}$, which is the subset of ultrafilters in X that do *not* contain a. (Recall that each ultrafilter contains either a or a', but not both). h is an isomorphism, because every non-zero element $a \in B$ is contained in an ultrafilter, and hence $h(a) = 0$ only if $a = 0$.

Evidently, $\mathscr{F} = h(\mathscr{B})$ is a field of subsets of X. $\mathscr{F}$ is a reduced field because if Φ_1 and Φ_2 are distinct elements of the set of ultrafilters X, then there is an $a \in \mathscr{B}$ such that $a \in \Phi_1$, $a \notin \Phi_2$ and so there is a subset of ultrafilters $h(a)$ containing Φ_1 but not Φ_2 (i.e. for each pair of points Φ_1, Φ_2 in X, there is a set in the field $\mathscr{F}$ containing Φ_1, but not Φ_2). $\mathscr{F}$ is perfect because h is an isomorphism between $\mathscr{F}$ and $\mathscr{B}$, and so each ultrafilter $\{h(a), h(b), \ldots\}$ in $\mathscr{F}$ corresponds to an ultrafilter $\{a, b, \ldots\}$ in $\mathscr{B}$. Thus, each $\mathscr{F}$-ultrafilter consists of those sets in $\mathscr{F}$ which contain a specific $\mathscr{B}$-ultrafilter. But the points of the set X are just the ultrafilters in $\mathscr{B}$, and so each ultrafilter in $\mathscr{F}$ is determined by a point in X.

The core of this theorem is the lemma that every non-zero element of a Boolean algebra is contained in an ultrafilter, i.e. it is the *existence* of this ultrafilter, guaranteed by the Axiom of Choice in the form of Zorn's Lemma, that is crucial for the existence of the isomorphism between the Boolean algebra and the field of sets.

The completeness theorem for the classical propositional calculus states that a sentence is consistent if and only if it is satisfiable. In the Lindenbaum-Tarski algebra $\mathscr{L}$, which is Boolean, this amounts to the statement that $|s| \neq 0$ if and only if there exists an interpretation that satisfies s. Now, every non-zero element $|s|$ is mapped by the Stone isomorphism onto a *non-empty* subset $h(|s|)$ in the field of sets that is isomorphic to $\mathscr{L}$. And every point in the subset $h(|s|)$ corresponds to an ultrafilter in $\mathscr{L}$ containing $|s|$, hence to a 2-valued homomorphism that maps $|s|$ onto 1, i.e. to an interpretation that satisfies s.

III. MECHANICS

The phase space of a classical mechanical system, say a free particle, is a Euclidean space, X, parametrized by the position and momentum coordinates of the particle. The physical magnitudes are real-valued functions on X, forming a commutative algebra. The idempotent magnitudes are represented by the characteristic functions on the Borel subsets of X. The characteristic function on the set Y, say E, is defined by

$$E(x) = 1 \quad \text{if} \quad x \in Y$$
$$E(x) = 0 \quad \text{if} \quad x \notin Y$$

The field $\mathscr{F}$ of Borel subsets of X under the partial ordering defined by set-inclusion is a complemented distributive lattice, a Boolean algebra. The Boolean algebra of idempotent magnitudes (propositions) is isomorphic to the Boolean algebra of Borel subsets of X, representing the possible events open to the system. The singleton subsets $\{x\} \in \mathscr{F}$ are *atoms* in $\mathscr{F}$. An atom in a Boolean algebra $\mathscr{B}$ is a minimal non-zero element of $\mathscr{B}$, i.e. an element $a \in \mathscr{B}$ such that there is no element of $\mathscr{B}$ between 0 and a. More precisely, a is an atom if and only if $a \neq 0$ and $b \leq a$ only if $b = 0$ or $b = a$.

It follows that a is an atom if and only if the principal filter generated by a is an ultrafilter. (For any $b \neq 0$, $a \wedge b \leq a$, and so if a is an atom either $a \wedge b = 0$ or $a \wedge b = a$. If $a \wedge b = a$, then $a \leq a \wedge b \leq b$. If $a \wedge b = 0$, then

$$a = a \wedge (b \vee b')$$
$$= (a \wedge b) \vee (a \wedge b')$$
$$= 0 \vee (a \wedge b')$$
$$= a \wedge b'$$

i.e.

$$a \leq b'.$$

So, for any $b \neq 0$, either b is a member of the principal filter Φ_a generated by a, or b' is a member of Φ_a. Hence Φ_a is an ultrafilter. Conversely, if Φ_a is an ultrafilter, for any $b \neq 0$, either $a \leqslant b$ or $a \leqslant b'$, If $c \leqslant a$ then either $a \leqslant c$, i.e. $c = a$, or $a \leqslant c'$, i.e. $c \leqslant a'$, in which case $c \leqslant a \wedge a'$, i.e. $c = 0$. So a is an atom.)

Equivalently, $a \in \mathscr{B}$ is an atom if and only if there is only one ultrafilter containing a. (Every ultrafilter containing an element $a \in \mathscr{B}$ includes the principal filter Φ_a generated by a. If a is an atom, Φ_a is an ultrafilter, and so is not included in any proper filter. If a is an atom, Φ_a is the only ultrafilter containing a. Conversely, if Φ is the only ultrafilter containing a, then for any non-zero $b \leqslant a$, Φ is the only ultrafilter containing b, because any filter containing b necessarily contains a. Hence $b = a$, because if a and b are distinct elements of $\mathscr{B}$ there exists an ultrafilter containing a but not b.)

Thus, there is a one-one correspondence between points in X and maximal consistent sets of propositions, i.e. with ultrafilters in the Boolean algebra of propositions. This is the significance of the classical mechanical notion of a *state*, the specification of a point in X. A state corresponds to an atom in $\mathscr{F}$, hence to an ultrafilter in the Boolean algebra of propositions, or a 2-valued homomorphism on this algebra, i.e. a Boolean assignment of truth values to the propositions.

In this sense, the phase space of a classical mechanical system is a topological characterization of the propositional structure: the phase space is the Stone space of the Boolean algebra of propositions. Under the Stone isomorphism, the image of a consistent set of propositions (i.e. a proper filter) is a non-empty closed subset in $\mathscr{F}$, and an ultrafilter corresponds to a singleton subset in $\mathscr{F}$. The unit filter is associated with the whole space, and the dual of the unit filter, the zero ideal, with the empty set.

I propose to view the transition from classical to quantum mechanics as involving the generalization of the Boolean event structures or propositional structures of classical mechanics to a particular class of non-Boolean structures. The propositional structure of a quantum mechanical system is a partial Boolean algebra isomorphic to the partial Boolean algebra of subspaces of a Hilbert space. The magnitudes of the system are represented by the self-adjoint operators on the Hilbert space, the idempotent magnitudes by projection operators. An atomic event is represented by an ultrafilter in the partial Boolean algebra, i.e. by a projection operator whose range is a 1-dimensional subspace or ray in the Hilbert space. The theorem of Kochen and Specker shows that the propositional structure of a quantum mechanical system – the logical structure of all possible events associated with a quantum mechanical system – is a partial Boolean algebra that is strongly non-Boolean, i.e. not imbeddable in a Boolean algebra.

To put this another way: The transition from classical to quantum mechanics involves a generalization of the classical notion of validity. The class of models over which validity is defined is extended to include partial Boolean algebras which are not imbeddable into Boolean algebras. I understand these models as representing possibility structures of events.

CHAPTER IX

IMBEDDABILITY AND VALIDITY

A classical propositional function $\varphi(x_1, \ldots, x_n)$ is a proposition-valued map on a classical propositional logic L. (Here the variables $x_1, \ldots, x_n$ range over propositions in L, not over individuals in a domain of interpretation as in the previous chapter.) To say that a particular propositional function, e.g. the function

$$x_1 \wedge (x_2 \wedge x_3) \equiv (x_1 \wedge x_2) \wedge x_3$$

is a classical tautology, is to say that every classical interpretation satisfies φ, whatever propositions are substituted for the variables x_1, x_2, x_3.

Now, an interpretation may be regarded as a 2-valued homomorphism from the Lindenbaum-Tarski algebra, $\mathscr{L}$, onto the 2-element Boolean algebra $\mathscr{Z}_2$. The validity of this propositional function may be expressed as the validity of the corresponding Boolean function

$$\varphi(x_1, x_2, x_3) = (\psi_1 \wedge \psi_2')' \wedge (\psi_2 \wedge \psi_1')'$$

where

$$\psi_1 = x_1 \wedge (x_2 \wedge x_3)$$
$$\psi_2 = (x_1 \wedge x_2) \wedge x_3.$$

(Recall the definition of the biconditional $\equiv$ in terms of conjunction and negation, and the correspondence between conjunction and infimum, and negation and complement. Here the variables x_1, x_2, x_3 range over elements in $\mathscr{L}$, and the symbol $\wedge$ denotes the infimum in $\mathscr{L}$.)

To say that the Boolean function is valid, is to say that the image or value of this function is mapped onto the unit element in $\mathscr{Z}_2$ by every 2-valued homomorphism, whatever elements in $\mathscr{L}$ are substituted for the variables. This amounts to saying that the function takes on the value 1 in $\mathscr{Z}_2$ for all substitutions of sequences from $\mathscr{Z}_2^3$ for the variables x_1, x_2 x_3. From an algebraic point of view, then, the notion of a classical tautology applies to a Boolean function $\varphi(x_1, \ldots, x_n)$, which takes on the value 1 in $\mathscr{Z}_2$ for all possible substitutions of sequences from $\mathscr{Z}_2^n$ for the variables.

Clearly, if φ is valid, and $\mathscr{B}$ is any Boolean algebra, then φ takes on the value 1 in $\mathscr{B}$ for all substitutions of elements from $\mathscr{B}$. To see this, consider a particular Boolean function, say the function $\varphi(x_1, x_2, x_3)$ above. Suppose

$$\varphi \neq 1 \quad \text{in} \quad \mathscr{B}$$

for some sequence from $\mathscr{B}^3$ substituted for the variables x_1, x_2, x_3, i.e.

$$\varphi(a_1, a_2, a_3) = a \neq 1 \quad \text{in} \quad \mathscr{B},$$

then there exists a 2-valued homomorphism, $h:\mathscr{B} \to \mathscr{L}_2$, such that

$$h(a) = 0 \quad \text{in} \quad \mathscr{L}_2,$$

i.e. $(h(a_1), h(a_2), h(a_3))$ is a sequence from $\mathscr{L}_2^3$ which yields the value 0 in $\mathscr{L}_2$ for the function $\varphi(x_1, x_2, x_3)$, which is impossible if φ is valid.

Moreover, the classical tautologies specify all Boolean functions which yield the unit element in a Boolean algebra when elements of the algebra are substituted for the variables – there are no other Boolean functions which are valid in this sense. One might regard the classical notion of validity as a logical concept restricted to possibility structures of events which are represented as Boolean algebras. (In the case of the predicate calculus, these Boolean algebras are set-algebras.) We seek a generalization of this notion to include possibility structures of events represented as partial Boolean algebras. Kochen and Specker propose that a propositional function such as the above is *valid in a partial Boolean algebra* $\mathscr{A}$ if every 'meaningful' substitution of elements from $\mathscr{A}$ into the associated Boolean function yields the unit element in $\mathscr{A}$. A 'meaningful' substitution is one which satisfies the compatibility relations; otherwise the partial operations are undefined in $\mathscr{A}$. In the above example, the elements a_1, a_2, a_3 of $\mathscr{A}$ substituted for x_1, x_2, x_3 are required to satisfy the conditions:

$$a_2 \leftrightarrow a_3$$
$$a_1 \leftrightarrow a_3$$
$$a_1 \leftrightarrow a_2 \wedge a_3$$
$$a_1 \wedge a_2 \leftrightarrow a_3$$
$$a_1 \wedge (a_2 \wedge a_3) \leftrightarrow (a_1 \wedge a_2) \wedge a_3.$$

This notion is formalized in the following definition: Let $a = (a_1, ..., a_n)$ be an element in $\mathscr{A}^n$, the n-fold Cartesian product $\mathscr{A} \times \mathscr{A} \times ... \mathscr{A}$ of the partial Boolean algebra $\mathscr{A}$. The domain, D_φ, in $\mathscr{A}$ of a Boolean function

$\varphi(x_1, ..., x_n)$ – regarded as a polynomial over $\mathscr{Z}_2$ – is defined recursively, together with a recursive definition of a map φ^* (corresponding to φ) from D_φ into $\mathscr{A}$, as follows:

   (1)   if φ is the polynomial 1, then $D_\varphi = \mathscr{A}^n$ and $\varphi^*(a) = 1$

   (2)   if φ is the polynomial x_i $(i = 1, ..., n)$, then $D_\varphi = \mathscr{A}^n$, and $\varphi^*(a) = a_i$

   (3)   if $\varphi = \psi + \chi$ or $\varphi = \psi \cdot \chi$ then D_φ consists of those sequences a which belong to the intersection of the domains of ψ and χ (i.e. $a \in D_\psi \cap D_\chi$), and also satisfy the compatibility condition $\psi^*(a) \leftrightarrow \chi^*(a)$.

The map $\varphi^*(a)$ is defined by $\varphi^*(a) = \psi^*(a) + \chi^*(a)$ or $\varphi^*(a) = \psi^*(a) \cdot \chi^*(a)$, respectively.

The definition of the domain of a Boolean function in a given partial Boolean algebra $\mathscr{A}$ serves to make precise the notion of a 'meaningful' substitution, while the map φ^* defines the value of the polynomial in $\mathscr{A}$ for each substitution.

The statement that the identity

$$\varphi(x_1, ..., x_n) = 1$$

holds in $\mathscr{A}$ is to be understood in the sense that

$$\varphi^*(a) = 1$$

for all $a \in D_\varphi$.

The statement that the identity

$$\varphi(x_1, ..., x_n) = \psi(x_1, ..., x_n)$$

holds in $\mathscr{A}$ is to be understood in the sense that

$$\varphi^*(a) = \psi^*(a)$$

for all $a \in D_\varphi \cap D_\psi$.

Now, the generalized definition of validity is this:

A propositional function $\varphi(x_1, ..., x_n)$ is *valid in the partial Boolean algebra* $\mathscr{A}$ if the identity $\varphi = 1$ holds in $\mathscr{A}$ for the corresponding Boolean function.

φ is *refutable in* $\mathscr{A}$ if for some $a \in D_\varphi$, $\varphi^*(a) = 0$ in $\mathscr{A}$.

φ is logically valid in the generalized sense, i.e. *Q-valid*, if φ is valid in every partial Boolean algebra $\mathscr{A}$.

If the choice of $\mathscr{A}$ is restricted to Boolean algebras, this definition of validity coincides with the usual definition: the set of valid propositional functions is just the set of classical tautologies. Thus the recursive definition of the domain of a propositional function coupled with the recursive definition of the map φ^* is a straightforward generalization of the classical, Boolean notion of satisfaction for φ.

It is important to appreciate the distinction between the validity of a propositional function

$$\psi \equiv \chi$$

in a partial Boolean algebra $\mathscr{A}$, and the holding of the identity

$$\psi = \chi$$

in $\mathscr{A}$. To say that $\psi \equiv \chi$ is valid in $\mathscr{A}$ is to say that

$$(\psi \equiv \chi) = 1$$

in $\mathscr{A}$, i.e. writing $\varphi = (\psi \equiv \chi)$ we require that

$$\varphi^*(a) = 1$$

for every sequence $a \in D_\psi \cap D_\chi$ satisfying the additional compatibility condition

$$\psi^*(a) \leftrightarrow \chi^*(a).$$

But, for the identity $\psi = \chi$ to hold in $\mathscr{A}$, we require that $\psi^*(a) = \chi^*(a)$ for *every* sequence $a \in D_\psi \cap D_\chi$, not only those sequences satisfying the additional compatibility condition $\psi^*(a) \leftrightarrow \chi^*(a)$. Thus, the set of admissable sequences $a \in \mathscr{A}^n$ is *smaller* in the case of the validity of the biconditional than in the case of the identity. If the identity holds in $\mathscr{A}$, then certainly the biconditional is valid in $\mathscr{A}$, but the converse is not in general true. The validity of the biconditional amounts to the holding of the identity for the restricted set of sequences which satisfy the compatibility condition $\psi^*(a) \leftrightarrow \chi^*(a)$.

For example, let $\varphi = (\psi \equiv \chi)$ be the classical tautology:

$$x_1 \wedge (x_2 \vee x_3) \equiv (x_1 \wedge x_2) \vee (x_1 \wedge x_3).$$

φ is not only valid in every partial Boolean algebra, it is also the case that the identity

$$x_1 \wedge (x_2 \vee x_3) = (x_1 \wedge x_2) \vee (x_1 \wedge x_3)$$

holds in every $\mathscr{A}$. For if $a = (a_1, a_2, a_3) \in D_\varphi$,

$$a_2 \leftrightarrow a_3$$
$$a_1 \leftrightarrow a_2$$
$$a_1 \leftrightarrow a_3.$$

But then a_1, a_2, a_3 generate a Boolean algebra. It follows that

$$a_1 \wedge (a_2 \vee a_3) = (a_1 \wedge a_2) \vee (a_1 \wedge a_3),$$

and hence

$$a_1 \wedge (a_2 \vee a_3) \leftrightarrow (a_1 \wedge a_2) \vee (a_1 \wedge a_3).$$

Thus, every sequence $a \in D_\psi \cap D_\chi$ automatically satisfies the compatibility condition $\psi^*(a) \leftrightarrow \chi^*(a)$.

In the case of a partial Boolean algebra $\mathscr{A}$ *imbeddable* into a Boolean algebra, the validity of the biconditional $\psi \equiv \chi$ in $\mathscr{L}_2$ (i.e. the classical tautologousness of the biconditional) entails the holding of the identity $\psi = \chi$ in $\mathscr{A}$. Thus, in the case of imbeddability (and only in this case): $\psi \equiv \chi$ is valid in $\mathscr{L}_2$ (and hence, as it turns out, valid in $\mathscr{A}$) is equivalent to $\psi = \chi$ holds in $\mathscr{A}$.

This follows from a theorem of Kochen and Specker, which establishes a relationship between the validity of classical tautologies in a partial Boolean algebra $\mathscr{A}$ and the imbeddability of $\mathscr{A}$ into a Boolean algebra. The statement of the theorem is as follows:

(1) A necessary and sufficient condition for the *imbeddability* of a partial Boolean algebra $\mathscr{A}$ into a Boolean algebra is the holding of the corresponding identity $\psi = \chi$ in $\mathscr{A}$ for every classical tautology of the form $\psi \equiv \chi$.

(2) A necessary and sufficient condition for the *weak imbeddability* of a partial Boolean algebra $\mathscr{A}$ into a Boolean algebra is the validity in $\mathscr{A}$ of every classical tautology.

(3) A necessary and sufficient condition for the existence of a *homomorphism* from a partial Boolean algebra $\mathscr{A}$ into a Boolean algebra is the irrefutability in $\mathscr{A}$ of every classical tautology.

A weak imbedding is a homomorphism which is an imbedding on Boolean subalgebras of $\mathscr{A}$. More precisely, a homomorphism h of $\mathscr{A}$ into $\mathscr{A}'$ is a weak imbedding if $h(a) \neq h(b)$ whenever $a \leftrightarrow b$ and $a \neq b$ in $\mathscr{A}$. Recall that a necessary and sufficient condition for the imbeddability of a partial Boolean algebra $\mathscr{A}$ into a Boolean algebra $\mathscr{B}$ is that for every pair of distinct elements $a, b \in \mathscr{A}$ there exists a homomorphism $h: \mathscr{A} \to \mathscr{L}_2$ which separates them in $\mathscr{L}_2$, i.e. such that $h(a) \neq h(b)$ in $\mathscr{L}_2$. Kochen and Specker label this important lemma Theorem 0. This result depends on the semi-simplicity property of Boolean algebras, i.e. essentially the homomorphism or ultrafilter theorem. The counterpart of Theorem 0 for weak imbeddability is the following: A necessary and sufficient condition for the weak imbeddability of a partial Boolean algebra $\mathscr{A}$ into a Boolean algebra $\mathscr{B}$ is that for every non-zero element $a \in \mathscr{A}$ there exists a homomorphism $h: \mathscr{A} \to \mathscr{L}_2$ such that $h(a) \neq 0$.

The first part of the theorem states that $\mathscr{A}$ is imbeddable into a Boolean algebra if and only if, for every propositional function of the form $\psi \equiv \chi$ which is *valid in* $\mathscr{L}_2$ (i.e. for which the identity $(\psi \equiv \chi) = 1$ holds in $\mathscr{L}_2$), the identity $\psi = \chi$ holds in $\mathscr{A}$.

Clearly, if $\mathscr{A}$ is imbeddable into a Boolean algebra $\mathscr{B}$, *all* the classical tautologies (i.e. all functions φ which are valid in $\mathscr{L}_2$) are valid in $\mathscr{A}$. For if φ is a classical tautology, φ is valid in $\mathscr{B}$, and hence certainly valid in $\mathscr{A}$. Validity in $\mathscr{B}$ requires $\varphi = 1$ in $\mathscr{B}$ for all sequences in B^n, while validity in $\mathscr{A}$ requires the holding of the identity only for those sequences in the domain of φ. Recall that the imbeddability of $\mathscr{A}$ into $\mathscr{B}$ means the existence of a one-one map into a part of $\mathscr{B}$.

The difference between weak imbeddability and (strong) imbeddability for the set of functions valid in $\mathscr{A}$ is just this: In the case of weak imbeddability all the classical tautologies are valid in $\mathscr{A}$ (and in general there are also functions valid in $\mathscr{A}$ which are not classical tautologies). In the case of (strong) imbeddability, all the classical tautologies are valid in $\mathscr{A}$. There may also be functions valid in $\mathscr{A}$ which are not classical tautologies. But here we know in addition that if $\psi \equiv \chi$ is a classical tautology, then $\psi = \chi$ holds in $\mathscr{A}$.

Thus, for weak imbeddability, if $\psi \equiv \chi$ is a classical tautology (i.e. if $\psi \equiv \chi$ is valid in $\mathscr{L}_2$), we know that $\psi \equiv \chi$ is valid in $\mathscr{A}$ (by the second part of the theorem), but we cannot conclude that $\psi = \chi$ holds in $\mathscr{A}$. In the case of (strong) imbeddability, this inference in legitimate, i.e. from the validity of a biconditional in $\mathscr{L}_2$, we may infer that the corresponding identity holds in $\mathcal{A}$. This means that in the case of imbeddability we may infer the holding of the identity

$$\psi = \chi \quad \text{in} \quad \mathcal{A}$$

from the validity of the biconditional

$$(\psi \equiv \chi) = 1 \quad \text{in} \quad \mathcal{A}$$

whenever $\psi \equiv \chi$ is a classical tautology, as well as the converse (which follows immediately from the definition of validity and identity).

Notice that we cannot conclude that *only* the classical tautologies are valid in $\mathcal{A}$ if $\mathcal{A}$ is imbeddable into a Boolean algebra: it does not follow that if $\mathcal{A}$ is imbeddable, and φ is valid in $\mathcal{A}$, then φ is valid in $\mathcal{L}_2$. For, to say that φ is valid in $\mathcal{A}$ is to say that

$$\varphi^*(a) = 1 \quad \text{in} \quad \mathcal{A}$$

for every $a \in D_\varphi$ in A^n, and to say that φ is valid in $\mathcal{L}_2$ is to say that

$$\varphi^*(a) = 1 \quad \text{in} \quad \mathcal{L}_2$$

for every a in $\mathcal{L}_2^n$. Now, if φ is valid in $\mathcal{L}_2$, i.e. a classical tautology, then φ is valid in every Boolean algebra, $\mathcal{B}$. But the Boolean imbeddability of $\mathcal{A}$ cannot guarantee that φ is valid in every Boolean algebra if φ is valid in $\mathcal{A}$. For this to follow it would be necessary – at least – that $\varphi^*(a) = 1$ in $\mathcal{A}$ for every $a \in \mathcal{A}^n$, not only for $a \in D_\varphi$, since a Boolean imbedding is a one-one homomorphism *into* $\mathcal{B}$, and all sequences $a \in \mathcal{B}^n$ are used in determining validity.

The necessity of the condition is relatively easy to prove. It is required to prove that the holding of the corresponding identity $\psi = \chi$ in $\mathcal{A}$ for every classical tautology of the form $\psi \equiv \chi$ is a necessary condition for the imbeddability of $\mathcal{A}$ into a Boolean algebra. In other words, if $\mathcal{A}$ is imbeddable, then for every biconditional which is a classical tautology (i.e. which is valid in $\mathcal{L}_2$), the corresponding identity $\psi = \chi$ holds in $\mathcal{A}$.

Suppose $\mathcal{A}$ is imbeddable into a Boolean algebra, and that $\psi \equiv \chi$ is a classical tautology. I shall show that this entails that the identity $\psi = \chi$ holds in $\mathcal{A}$ by proving that the converse

$$\psi \neq \chi \quad \text{in} \quad \mathcal{A}$$

leads to a contradiction.

If $\psi \neq \chi$ in $\mathscr{A}$, then for some $a \in D_\psi \cap D_\chi$

$$\psi^*(a) \neq \chi^*(a)$$

Now, by Theorem 0, since $\mathscr{A}$ is imbeddable into a Boolean algebra, for each b, $c \in \mathscr{A}$ $(b \neq c)$ there exists a homomorphism $h : \mathscr{A} \to \mathscr{L}_2$ such that

$$h(b) \neq h(c)$$

and so there exists a homomorphism $h : \mathscr{A} \to \mathscr{L}_2$ such that

$$h(\psi^*(a)) \neq h(\chi^*(a))$$

or

$$\psi^*(h(a_1), ..., h(a_n)) \neq \chi^*(h(a_1), ..., h(a_n)).$$

In other words, $(h(a_1), ..., h(a_n))$ is a sequence in $\mathscr{L}_2^n$ such that

$$\psi^*(h(a_1), ..., h(a_n)) \neq \chi^*(h(a_1), ..., h(a_n)).$$

This means that $\psi = \chi$ in Z_2, and so the biconditional $\psi \equiv \chi$ is not valid in $\mathscr{L}_2$, i.e. $\psi \equiv \chi$ is not a classical tautology, contrary to our original assumption.

Notice that it would not in general be permissable to infer the non-validity of the biconditional $\psi \equiv \chi$ from the fact that the identity $\psi = \chi$ fails to hold in a partial Boolean algebra. This inference is, however, obviously legitimate in $\mathscr{L}_2$.

To prove the sufficiency of the condition, it is necessary to show that the holding of the corresponding identity $\psi = \chi$ in $\mathscr{A}$ for every classical tautology of the form $\psi \equiv \chi$ entails the Boolean imbeddability of $\mathscr{A}$. Kochen and Specker prove the contrapositive: If $\mathscr{A}$ is not imbeddable into a Boolean algebra, then there exists a classical tautology $\psi \equiv \chi$ such that $\psi \neq \chi$ in $\mathscr{A}$.

They consider the set of sentences K_1, formulated in some first-order language L, describing all equations of the form $\alpha + \beta = \gamma$ or $\xi \eta = \zeta$ which hold among elements of $\mathscr{A}$, together with the set of axioms, K_2, characterizing the class of Boolean algebras. Thus, the class of all models of the set of sentences $K = K_1 \cup K_2$ comprises all homomorphic images of $\mathscr{A}$ which are Boolean algebras.

Now, if $\mathscr{A}$ is *not* imbeddable into a Boolean algebra, then, by Theorem 0, there exists a pair of elements a, $b \in \mathscr{A}$ such that *no homomorphism into $\mathscr{L}_2$ will separate them*. That is, a and b are two distinct elements in $\mathscr{A}$

which are identified by every homomorphism into $\mathscr{L}_2$. If a and b are not separated by any homomorphism into $\mathscr{L}_2$, then they cannot be separated by a homomorphism into *any* Boolean algebra (by the ultrafilter theorem, or the semi-simplicity property of Boolean algebras). That it so say, *a and b are identified in every model of K* (since the models of K are just a class of Boolean algebras, viz. those which are homomorphic images of $\mathscr{A}$).

Kochen and Specker construct a formula in L which is provable from K_2, the set of axioms for a Boolean algebra. The formula says, in effect, that two distinct elements in $\mathscr{A}$ are identified if some finite set of relations of the form $\alpha + \beta = \gamma$ and $\xi\eta = \zeta$ hold in $\mathscr{A}$. This formula is therefore valid in all Boolean algebras. If ψ is the Boolean function corresponding to the conditional $x \supset \varrho$, where ϱ is the formula in L describing the set of relations $\alpha + \beta = \gamma$ and $\xi\eta = \zeta$, and χ is the Boolean function corresponding to the conditional $y \supset \varrho$, with x and y variables in L ranging over elements in $\mathscr{A}$. it follows that

$$\psi = \chi$$

holds in $\mathscr{L}_2$ and hence that

$$\psi \equiv \chi$$

is valid in $\mathscr{L}_2$, i.e. that $\psi \equiv \chi$ is a classical tautology. But $\psi = \chi$ does not hold in $\mathscr{A}$, by construction.

Thus, in the case of a partial Boolean algebra for which there is no Boolean imbedding, a biconditional is constructed which is a classical tautology, but for which the corresponding identity, $\psi = \chi$, does not hold in $\mathscr{A}$.

The relation between imbeddability and classical validity established by this theorem shows that there is a propositional function φ which is a classical tautology but which is not Q-valid. In particular, there is a propositional function φ which is not valid in the partial Boolean algebra $\mathscr{A}_3$ isomorphic to the partial Boolean algebra of subspaces of $\mathscr{H}_3$, since there is no 2-valued homomorphism on $\mathscr{A}_3$, and hence no imbedding of $\mathscr{A}_3$ into a Boolean algebra. That is to say, there is a Boolean function $\varphi(x_1, ..., x_n)$ which maps every sequence $(a_1, ..., a_n)$ in $\mathscr{L}_2^n$ onto the unit in $\mathscr{L}_2$, but which does not map every sequence in the domain of φ in $\mathscr{A}_3^n$ onto the unit in $\mathscr{A}_3$.

Kochen and Specker construct the following example: Consider the

Boolean function

$$\psi = \prod(x_i + x_j + x_k + x_i x_j x_k)$$

where each factor in the product, $x_i + x_j + x_k + x_i x_j x_k$, coresponds to an orthogonal triple of vectors in the set of 117 vectors used by Kochen and Specker in their imbeddability theorem. Each factor takes on the value 1 in $\mathcal{L}_2$ only for sequences (a_1, a_2, a_3) with exactly one a_i equal to 1. The non-existence of a 2-valued homomorphism on the finite partial Boolean sub-algebra generated by the 117 vectors means that there is no sequence of 1's and 0's such that one 1 and two 0's is substituted for the variables x_i, x_j, x_k in each factor $x_i + x_j + x_k + x_i x_j x_k$. Hence ψ takes on the value 0 in $\mathcal{L}_2$ for *every* sequence in $\mathcal{L}_2^n$. It follows that the function

$$\varphi = 1 - \prod(x_i + x_j + x_k + x_i x_j x_k)$$

is classically valid – takes the value 1 in $\mathcal{L}_2$ for every sequence in $\mathcal{L}_2^n$.

Now the Boolean function

$$x_i \vee x_j \vee x_k \quad \text{i.e.} \quad x_i + x_j + x_k + x_i x_j x_k - x_i x_j - x_i x_k - x_j x_k$$

maps any sequence (a_1, a_2, a_3) in $\mathcal{A}_3^3$ corresponding to three mutually orthogonal subspaces in $\mathcal{H}_3$ onto the unit element in $\mathcal{A}_3$, and so the function $x_i + x_j + x_k + x_i x_j x_k$ maps every sequence onto the unit. Clearly, then, there exists a sequence in $\mathcal{A}_3^n$ which the function ψ maps onto the unit in $\mathcal{A}_3$, and hence there exists a sequence which φ maps onto 0. Thus, φ is a classical tautology which is not valid in $\mathcal{A}_3$.

This theorem puts the significance of the non-imbeddability of the propositional structure of quantum mechanics into a Boolean algebra in a new light. If the valid propositional functions are regarded as invariants characterizing the logical structure of events – just as the invariants of a group of geometrical transformations characterize the space-time structure of events – then the Boolean imbeddability of the partial Boolean algebra of sub-spaces of $\mathcal{H}_2$ means that this logical structure is essentially Boolean: all the classical tautologies are valid in A_2 (as well as some propositional functions which are not classical tautologies). The classical tautologies will remain valid under an extension of the structure to a Boolean structure $\mathcal{B}$, while those propositional functions which are valid in $\mathcal{A}_2$ but not classical tautologies will no longer map all sequences in $\mathcal{B}$ onto the unit.

But in the case of logical structures $\mathscr{A}_3$, $\mathscr{A}_3$, ..., generated by the subspaces of $\mathscr{H}_3$, $\mathscr{H}_4$, ..., some classical tautologies are not valid. The structures are essentially non-Boolean: there is no possible extension of $\mathscr{A}_3$ to a Boolean algebra $\mathscr{B}$, because *all* classical tautologies are valid in $\mathscr{B}$, yet some propositional functions which are classical tautologies do not map every sequence in $\mathscr{A}_3^n$ onto the unit in $\mathscr{A}_3$.

CHAPTER X

THE STATISTICS OF NON-BOOLEAN
EVENT STRUCTURES

The Kochen and Specker theory of partial Boolean algebras leads to the resolution of the core problem of interpretation of quantum mechanics, the problem of hidden variables. To recapitulate: Quantum mechanics incorporates an algorithm for assigning probabilities to ranges of values of the physical magnitudes:

$$p_W(a \in S) = \text{Tr}(W P_A(S))$$

where W represents a statistical state of the theory, and $P_A(S)$ is the projection operator onto the subspace in Hilbert space associated with the range S of the magnitude A. The statistical states generate all possible (generalized) probability measures on the partial Boolean algebra of subspaces of Hilbert space. Joint probabilities

$$p_W(a_1 \in S_1 \ \& \ a_2 \in S_2 \ \& \ \cdots \ \& \ a_n \in S_n) =$$
$$= \text{Tr}(W P_{A_1}(S_1) P_{A_2}(S_2) \ldots P_{A_n}(S_n))$$

are defined only for compatible magnitudes $A_1, A_2, \ldots, A_n$, and there are no dispersion-free statistical states. The problem of hidden variables concerns the possibility of representing the statistical states of quantum mechanics by measures on a classical probability space in such a way that the algebraic structure of the magnitudes of the theory is preserved. This is the problem of imbedding the partial algebra of magnitudes into a commutative algebra or, equivalently, the problem of imbedding the partial Boolean algebra of idempotent magnitudes (properties, propositions) into a Boolean algebra. The imbedding turns out to be impossible; there are no 2-valued homomorphisms on the partial Boolean algebra of idempotents of a quantum mechanical system, except in the case of a system associated with a 2-dimensional Hilbert space. Thus, the transition from classical to quantum mechanics involves the generalization of the Boolean propositional or event structures of classical mechanics to a particular class of non-Boolean structures. This may be understood as a generalization of the classical notion of validity: The class of models over which validity is defined is extended to include partial Boolean algebras which are not imbeddable into Boolean algebras.

In a Boolean algebra $\mathscr{B}$, there is a one-one correspondence between atoms, ultrafilters, and 2-valued homomorphisms, essentially because an ultrafilter Φ in $\mathscr{B}$ contains a or a', but not both, for every $a \in \mathscr{B}$. If b is an atom, either a or a' is above b, i.e. either $a \leqslant b$ or $a \leqslant b'$ for every $a \in \mathscr{B}$ (but not both, or else $b = 0$). Hence, there can be one and only one ultrafilter containing an atom. A 2-valued homomorphism is definable on $\mathscr{B}$ by mapping each element $a \in \mathscr{B}$ onto 1 or 0 according to whether a is or is not a member of the ultrafilter Φ.

In a partial Boolean algebra that is not imbeddable in a Boolean algebra, the one-one correspondence between atomic events, ultrafilters, and 2-valued homomorphisms no longer holds. The partial Boolean algebra may be regarded as a partially ordered system, so the notion of a filter (and hence an ultrafilter as a maximal filter) is still well-defined. But it is no longer the case that if Φ is an ultrafilter, then or each $a \in \mathscr{A}$ either $a \in \Phi$ or $a' \in \Phi$, and hence ultrafilters do not define 2-valued homomorphisms on $\mathscr{A}$. This is because ultrafilters (maximal filters) are no longer *prime* filters. A filter Φ is prime if it is proper (i.e. a proper subset of $\mathscr{A}$), and if $a \vee b \in \Phi$ only if either $a \in \Phi$ or $b \in \Phi$. Every ultrafilter Φ in a partial Boolean algebra contains the unit, and hence contains $a \vee a'$ for every $a \in A$. But if Φ is an ultrafilter in the maximal Boolean sub-algebra $\mathscr{B} \subset \mathscr{A}$, then neither a nor a' will belong to Φ if a and a' are outside $\mathscr{B}$, i.e. incompatible with the elements contained in Φ. An atom in $\mathscr{A}$ will correspond to an ultrafilter, but not to a prime filter, and hence will not define a 2-valued homomorphism on $\mathscr{A}$.

The Stone isomorphism maps every element in a Boolean algebra onto the set of ultrafilters containing the element. Thus, a measure on a classical probability space X may be interpreted as a measure over ultrafilters or atomic events in a Boolean algebra $\mathscr{B}$, the points $x \in X$ corresponding to ultrafilters in $\mathscr{B}$ and the singleton subsets $\{x\}$ in $\mathscr{F}$ corresponding to atomic events. The probability of an event a may be understood as the measure of the set of ultrafilters containing a, or the measure of the set of atomic events that can occur together with the event a:

$$p(a) = \mu(\Phi_a).$$

The conditional probability of a given b, $p(a \mid b)$, is the measure of the set of ultrafilters containing a in the set of ultrafilters containing b, with respect to a renormalized measure assigning probability 1 to the set Φ_b:

$$p(a \mid b) = \frac{\mu(\Phi_a \cap \Phi_b)}{\mu(\Phi_b)} = \frac{p(a \wedge b)}{p(b)}.$$

Loosely; we 'count' the number of atomic events that can occur together with the event b, in the set of atomic events that can occur together with the event a. Notice that if b is an atom, the conditional probability is a 2-valued measure.

The statistical states of quantum mechanics define probability measures in the classical sense on each maximal Boolean sub-algebra of the partial Boolean algebra of propositions of a quantum mechanical system. Consider a system associated with a 3-dimensional Hilbert space $\mathscr{H}_3$. Let A and B be two incompatible (non-degenerate) magnitudes with eigenvalues a_1, a_2, a_3 and b_1, b_2, b_3, respectively. The corresponding eigenvectors are α_1, α_2, α_3 and β_1, β_2, β_3. I shall also denote the atoms (atomic propositions or events) in the maximal Boolean subalgebras $\mathscr{B}_A$ and $\mathscr{B}_B$ of $\mathscr{A}$ by a_i and b_j, i.e., I shall use the same symbols to denote properties of the systems represented by these values of the magnitudes.

The statistical state associated with the vector α_1 assigns probabilities

$$P_{\alpha_1}(a_1) = 1, \qquad P_{\alpha_1}(a_2) = 0, \qquad P_{\alpha_1}(a_3) = 0$$

to the atomic propositions in $\mathscr{B}_A$, and probabilities $p_{\alpha_1}(b_1) = |(\beta_1, \alpha_1)|^2$, $p_{\alpha_1}(b_2) = |(\beta_2, \alpha_1)|^2$, $p_{\alpha_1}(b_3) = |(\beta_3, \alpha_1)|^2$ to the atomic propositions in $\mathscr{B}_B$. How are these probabilities to be understood? Since there are no 2-valued homomorphisms on $\mathscr{A}_3$, the probability $p_{\alpha_1}(b_1)$, for example, cannot be interpreted as the conditional probability, $p(b_1 \mid a_1)$, that the proposition b_1 is true (or the corresponding event obtains) given that the proposition a_1 is true, i.e. the probability that the value of the magnitude B is b_1 given that the value of the magnitude A is a_1.

The problem at issue is this: Suppose a system S has the property a_1. The statistical algorithm of quantum mechanics assigns (non-zero) probabilities to properties incompatible with a_1, for example $p_{\alpha_1}(b_1) = |(\beta_1, \alpha_1)|^2$. These probabilities cannot be understood as conditional probabilities. The probability assigned to b_1 by the statistical state α_1 cannot be interpreted as the relative measure of the set of ultrafilters containing b_1 in the set of ultrafilters containing a_1, because, firstly, a_1 and b_1

are atoms in $\mathscr{A}_3$ and, secondly, a_1 and b_1 cannot be represented as non-atomic propositions in a Boolean algebra because no Boolean imbedding of $\mathscr{A}_3$ is possible. What do these probabilities mean?

I shall show that the generalized measures on the partial Boolean algebras of quantum mechanics satisfy a law of large numbers in an analogous sense to the measures on a classical probability space or Boolean event structure.

Suppose $A^{(1)}$, $A^{(2)}, ..., A^{(n)}$ are independent random variables on a classical probability space X that are *statistically equivalent*, i.e.

$$p(a^{(i)} \in S) = \mu(A^{(i)-1}(S)) = f(S)$$

for all $A^{(i)}$, and all $S \in \mathscr{F}$, so that

$$\text{Exp}_\mu(A^{(i)}) = k$$

for all $A^{(i)}$. Here f is the 'distribution function' of the random variable. To say that the random variables are statistically equivalent, is to say that they all have the same distribution function. The symbol $a^{(i)}$ is a variable denoting a general value of the magnitude $A^{(i)}$, so $a^{(i)} \in S$ is to be read: the value of the magnitude $A^{(i)}$ lies in the range S.

Define the random variable

$$\bar{A} = \frac{1}{n}(A^{(1)} + A^{(2)} + \cdots + A^{(n)}).$$

Let Δk be a neighbourhood of k, i.e. an interval $(k-\delta, k+\delta)$ where $\delta > 0$. Then, it can be shown that for any Δk (i.e. for any $\delta > 0$)

$$p_\mu(\bar{a} \in \Delta k) \to 1$$

as $n \to \infty$. This is the classical law of large numbers. (The symbol $\bar{a}$ is a variable denoting a general value of the magnitude $\bar{A}$. Thus $\bar{a} \in \Delta k$ is to be read: the value of the magnitude $\bar{A}$ lies in the range Δk.)

To put this another way: Consider the n-fold Cartesian product of the space X

$$\bar{X} = X^{(1)} \times X^{(2)} \times \cdots \times X^{(n)}$$

with the probability measure

$$\bar{\mu} = \mu^{(1)} \times \mu^{(2)} \times \cdots \times \mu^{(n)}$$

where $\mu^{(i)}$ is the same probability on $X^{(i)}$ as $\mu^{(j)}$ is on $X^{(j)}$, the measure

μ. Define the random variable

$$\bar{A} = \frac{1}{n}(A^{(1)} + A^{(2)} + \cdots + A^{(n)})$$

where $A^{(i)}$ is the same random variable on $X^{(i)}$ as $A^{(j)}$ is on $X^{(j)}$, say A. (Actually, $\bar{A}$ should be defined as

$$\bar{A} = \frac{1}{n}(A^{(1)} \times I^{(2)} \times \cdots \times I^{(n)} + I^{(1)} \times A^{(2)} \times$$
$$\times I^{(3)} \times \cdots \times I^{(n)} + \cdots + I^{(1)} \times \cdots \times I^{(n-1)} \times A^{(n)})$$

where $I^{(i)}$ is the unit random variable – real-valued function – on the space $X^{(i)}$.) Let $\mathrm{Exp}_\mu(A) = k$. Then, for any Δk:

$$\bar{\mu}[\bar{A}^{-1}(\Delta k)] \to 1$$

as $n \to \infty$, i.e. the measure in $\bar{X}$ of the set of points assigning a value to the random variable $\bar{A}$ in an infinitesimally small range about $\mathrm{Exp}_\mu(A)$ tends to 1 as the number of factor spaces in $\bar{X}$ tends to infinity.

Loosely: If we understand the probability space X as exhibiting the possible events open to a certain physical system (i.e. the propositional structure of a system), and these events are weighted by a measure function determining the average or expectation value of a random variable A as $\mathrm{Exp}_\mu(A) = k$, then if we consider a very large number n of non-interacting copies of the system as a new composite system, in the limit as $n \to \infty$ the probability is 1 that the value of the magnitude $\bar{A}$ of the composite system is equal to k.

Now, the expectation value of an idempotent magnitude is the probability of the corresponding proposition or event. Thus, to say of a system that the probability of the proposition a corresponding to the idempotent magnitude P_a is $p_\mu(a)$, is to say that if we take n non-interacting copies of the system, then in the limit as $n \to \infty$ the probability is 1 that the value of the magnitude $\bar{P}_a$ of the composite system is equal to $p_\mu(a)$. In other words, *the probability tends to 1 that the composite system (i.e. the statistical ensemble) has the property $p_\mu(a)$, i.e. the property corresponding to this value of the magnitude $\bar{P}_a$.* And this property of the composite system is, loosely, the property that a fraction, $\mathrm{exp}_\mu(P_a) = p_\mu(a)$, of the component systems has the property a.

Notice that if the measure μ is a 2-valued measure assigning the probability 1 to the *atomic* proposition a, i.e. if the statistical ensemble is constituted of systems with the property a, then the probability in the ensemble tends to 1 that the value of the magnitude $\bar{P}_b$ for any idempotent $P_b \neq P_a$ is equal to 0. (For $\bar{P}_a$, of course, the value is 1.) In particular, the fraction of systems in the ensemble with an atomic property $b \neq a$ is zero.

In the case of a partial Boolean algebra like $\mathscr{A}_3$ the following theorem can be proved (see Finkelstein): Let $\psi^{(i)}$ be a unit vector in $\mathscr{H}_3^{(i)}$, generating the statistics of a pure statistical state, so that $\text{Exp}_{\psi_{(i)}}(A^{(i)}) = k$ for the magnitude $A^{(i)}$. Consider the vector

$$\bar{\psi} = \psi^{(1)} \otimes \psi^{(2)} \otimes \cdots \otimes \psi^{(n)}$$

in the tensor product Hilbert space

$$\mathscr{H}_3 = \mathscr{H}_3^{(1)} \otimes \mathscr{H}_3^{(2)} \otimes \cdots \otimes \mathscr{H}_3^{(n)}$$

where $\psi^{(i)}$ is the same vector in $\mathscr{H}_3^{(i)}$ as $\psi^{(j)}$ is in $\mathscr{H}_3^{(j)}$, say ψ, and the magnitude

$$\bar{A} = \frac{1}{n}(A^{(1)} + A^{(2)} + \cdots + A^{(n)})$$

where $A^{(i)}$ is the same operator in $\mathscr{H}_3^{(i)}$ as $A^{(j)}$ is in $\mathscr{H}_3^{(j)}$, say A. (Again, $\bar{A}$ should be defined as

$$\bar{A} = \frac{1}{n}(A^{(1)} \otimes I^{(2)} \otimes \cdots \otimes I^{(n)} + I^{(1)} \otimes A^{(2)} \otimes$$
$$\otimes I^{(3)} \otimes \cdots \otimes I^{(n)} + \cdots + I^{(1)} \otimes \cdots \otimes I^{(n-1)} \otimes A^{(n)})$$

where $I^{(i)}$ is the unit operator on $\mathscr{H}_3^{(i)}$.) Then $\bar{\psi}$ is practically an eigenvector of the operator $\bar{A}$ in $\mathscr{H}_3$, with the eigenvalue k, even though ψ is not an eigenvector of A in $\mathscr{H}_3$, i.e.

$$\lim_{n \to \infty} \|\bar{A}\bar{\psi} - k\bar{\psi}\| = 0.$$

This holds for any operator. What this theorem says is that if we consider n non-interacting copies of a system whose statistics is generated by a vector ψ as a new composite system, then the statistics of *this* system is generated by a vector which is practically an eigenvector of the operator representing the magnitude $\bar{A}$, with eigenvalue $\text{Exp}_\psi(A)$, if n is large, for any magnitude A and any vector ψ. Thus, if we take n non-interacting copies of a system for which the proposition a_1 is true, so that the statistics of the ensemble is generated by the vector α_1, then the above theorem applied to the idempotent magnitude P_{b_1} yields:

$$\lim_{n \to \infty} \|\bar{P}_{b_1}\bar{\alpha}_1 - k\bar{\alpha}_1\| = 0.$$

Loosely: in the limit as $n \to \infty$, $\bar{\alpha}_1$ is practically an eigenvector of the operator representing the magnitude $\bar{P}_{b_1}$ with eigenvalue k, where k is the expectation value of P_{b_1} specified by the vector α_1.

Now, the expectation value of an idempotent magnitude is the probability of the corresponding proposition, i.e. $\mathrm{Exp}_{\alpha_1}(P_{b_1}) = p_{\alpha_1}(b_1)$. And to say that $\bar{\alpha}_1$ is practically an eigenvector of the operator representing the magnitude $\bar{P}_{b_1}$ with eigenvalue k, is to say, in effect, that the probability is very close to 1 that the magnitude $\bar{P}_{b_1}$ takes the value k, i.e. in the limit as $n \to \infty$, the probability is 1 that the magnitude $\bar{P}_{b_1}$ takes the value $p_{\alpha_1}(b_1)$. (More precisely, for any neighbourhood Δk of k

$$p_{\bar{\alpha}_1}(\bar{p}_{b_1} \in \Delta k) \to 1$$

as $n \to \infty$, where $\bar{p}_{b_1}$ is a variable denoting a general value of the magnitude $\bar{P}_{b_1}$.) In other words, *the probability tends to 1 that the composite system (i.e. the ensemble) has the property $p_{\alpha_1}(b_1)$, i.e. the property corresponding to this value of the magnitude $\bar{P}_{b_1}$.* And this property of the composite system is, loosely, the property that a fraction, $\exp_{\alpha_1}(P_{b_1}) = p_{\alpha_1}(b_1)$, of the component systems have the property b_1.

Thus, a system S with the atomic property a_1 is associated with the statistical state α_1 in the sense that with respect to the property b_1, say, S is to be regarded as a member of a statistical ensemble considered as a composite system with the property $p_{\alpha_1}(b_1)$ corresponding, loosely, to a fraction $p_{\alpha_1}(b_1)$ of the component systems having the property b_1. But this is not to say that some component systems in the ensemble have the property $a_1 \wedge b_1$, for just as in the Boolean case, an atomic proposition is a maximal specification of the properties of a system, and no system is characterized by the conjunction of two atomic propositions, whether or not these are incompatible. The probabilities assigned to atoms in $\mathcal{B}_A$, $\mathcal{B}_B$, etc., are determined by the law of large numbers as properties of an ensemble homogeneous with respect to the atomic property a_1, but such an ensemble is not defined by a 2-valued measure on $\mathscr{A}_3$. In a Boolean propositional structure, the probabilities generated in this way are all 0. The assignment of non-zero probabilities in $\mathscr{A}_3$ to atoms incompatible with the designated atom is directly related to the non-trivial compatibility relation.

(It might be supposed that the non-existence of systems in the ensemble with the property $a_1 \wedge b_1$ depends on an arbitrary restriction of the conjunction operation to compatible elements. In a partial Boolean algebra, $a \wedge b$ is defined as an element in the algebra if and only if a and b are compatible. In the partial Boolean algebra of subspaces of a Hilbert space, $\mathscr{K}_a \wedge \mathscr{K}_b$ denotes the infimum of the subspaces $\mathscr{K}_a$ and $\mathscr{K}_b$, and this is *always* defined, even if the subspaces are incompatible: the subspaces of a Hilbert space actually form a lattice. Thus, conjunction in the sense of infimum is always defined. But this is quite irrelevant. There are no subspaces in the lattice which are left out of the partial Boolean algebra. The infimum and supremum of two incompatible subspaces $\mathscr{K}_a$, $\mathscr{K}_b$ are elements in the partial Boolean algebra as well, only these elements not related to $\mathscr{K}_a$ and $\mathscr{K}_b$ by the binary operations of the algebra.)

These considerations further clarify the way in which the properties of a quantum mechanical system hang together, and the difference between this propositional structure and the Boolean structure of the properties of a classical mechanical system. Since the propositional structure of a quantum mechanical system is not imbeddable in a Boolean algebra, there is no sense in which this structure is 'incomplete' relative to the propositional structures of classical systems. A quantum mechanical system has all its properties in the same sense in which a classical mechanical system has all its properties: the difference lies in the way in which these properties are structured. In the Boolean case there is a correspondence between atoms and 2-valued homomorphisms, hence 2-valued measures, and so an ensemble homogeneous in some atomic property is characterized by a 2-valued measure which selects an ultrafilter of propositions, i.e. assigns probabilities 0 or 1 to every range of every magnitude. In a partial Boolean algebra like $\mathscr{A}_3$, the algebraic relations between the incompatible atoms in $\mathscr{B}_A$ and $\mathscr{B}_B$ determine multi-valued measures $p_{\alpha_i}(b_j) = p_{\beta_j}(a_i)$. Since these measures satisfy a law of large numbers, they may be regarded as probabilities in the same sense as the 2-valued measures of classical mechanics. There should be no special problem concerning the meaning of the multi-valued probabilities assigned to the properties of a quantum mechanical system by the specification of an atomic property of the system. They mean whatever the 0, 1 probabilities assigned to the properties of a classical system by the specification of an atomic property of the system are understood to mean.

CHAPTER XI

THE MEASUREMENT PROBLEM

Underlying the disturbance theory of measurement of the Copenhagen interpretation is the assumption that the propositional structures of physical systems are Boolean algebras, i.e. that the properties of a physical system can hang together only in a Boolean structure. On this assumption, the peculiar statistical relations of quantum mechanics reflect either the incompleteness of the theory and the existence of hidden variables, or necessary measurement restrictions on measurements at the microlevel. By rejecting the first explanation, the Copenhagen interpretation treats the magnitudes of the theory as dispositional, in the sense that the probabilities refer to the dispositions for a microsystem to be 'disturbed' in certain ways in situations defined by measuring instruments. Formally, a measuring instrument selects a particular maximal Boolean sub-algebra and hence a particular set of possible disturbances for the system. Thus, the probability assignments generated by the statistical states are usually characterized as the probabilities of finding values for the magnitudes if appropriate measurements are made. A measurement is not understood as a procedure for establishing whether or not a system has a certain property, but as a disturbance of a certain kind characterized by the appropriate magnitude.

I have argued that von Neumann's basically analytical approach was to some extent undermined by an uncritical acceptance of certain aspects of the disturbance theory of measurement. He saw the non-existence of joint probability assignments to incompatible propositions as a feature of the theory directly related to the disturbance character of measurement at the microlevel. Since measurements disturb, and the probabilities represent dispositions for the system to be disturbed in certain ways in certain situations, the theory comprising the statistical algorithm and equation of motion alone would seem to be incomplete, since it does not determine the peculiar stochastic transitions that occur in measurement processes at the microlevel. To complete the theory, von Neumann proposed a measurement postulate that has since become known as the 'projection postulate'.

His argument for this postulate proceeds from an analysis of the experiment of Compton and Simon. The experiment involves a collision interaction between photons and electrons. Given the initial paths of the photon and electron before the collision, and the direction of the momentum transfer, the paths of the particles after collision are determined by the theory. The experiment may be regarded as providing a refutation of the Bohr-Kramers-Slater theory, a preliminary version of quantum mechanics in which energy and momentum are conserved only statistically and not in individual collision processes. If we assume the laws of collision (as given by the quantum theory) and take the paths of the particles before the collision as known, then measurement of the path of *either* the photon *or* the electron after the collision suffices to determine the direction of the momentum transfer. What the Compton-Simon experiment shows, von Neumann argues, is that these two measurements, M_1 and M_2, give the same result, i.e. the same physical magnitude A (the direction of the momentum transfer at the point of collision) is measured in two different ways (by detecting the photon, and by detecting the electron, after the collision) and the results always agree.

Now, the quantum statistical state associated with the state of motion of the photon and electron before the collision does not determine a unique direction of momentum transfer at the point of collision, but assigns probabilities to the possible values of this magnitude A. In the experiment, the time difference between the measurements M_1 and M_2 is usually of the order of 10^{-10} seconds. Prior to the measurement M_1 of the magnitude A, then, the result of the measurement is only statistically determined. But after the measurement M_1, the result of M_2 is uniquely determined. (Von Neumann, pp. 213, 214; I have altered von Neumann's symbols to conform with my notation.)

We can formulate the principle that is involved as follows: by nature, three degrees of causality or non-causality may be distinguished. First, the A value could be entirely statistical, i.e., the result of a measurement could be predicted only statistically; and if a second measurement were taken immediately after the first one, this would also have a dispersion, without regard to the value found initially – for example, its dispersion might be equal to the original one. Second, it is conceivable that the value of A may have a dispersion in the first measurement, but that immediately subsequent measurement is constrained to give a result which agrees with that of the first. Third, A could be determined causally at the outset.

The Compton-Simons (*sic*) experiment now shows that only the second case is possible in a statistical theory .Therefore, if the system is initially found in a state in which the values of A cannot be predicted with certainty, then this state is transformed by a measurement M of A (in the example above, M_1) into another state: namely, into one in which the value of A is uniquely determined. Moreover, the new state, in which M places the system, depends not only on the arrangement of M, but also on the result of the measurement M (which could not be predicted causally in the original state) – because the value of A in the new state must actually be equal to this M-result.

On this basis, von Neumann argues that if a measurement of the magnitude A with (non-degenerate) eigenvalues a_1, a_2, ... and corresponding eigenvectors α_1, α_2, ... yields the result a_1, then the initial statistical state is transformed to a new statistical state determined by the vector α_1. For only the statistical operator P_{α_1} represents the quantum statistical state assigning a probability of 1 to the value a_1 of A. (Von Neumann, p. 217. Again, I have altered von Neumann's symbols to conform with my notation.)

We have then answered the question as to what happens in the measurement of a quantity A, under the above assumptions for its operator A. To be sure the 'how' remains unexplained for the present. This discontinuous transition from ψ into one of the states α_1, α_2, ... (which are independent of ψ, because ψ enters only into the respective probabilities $p(a_i) = |(\alpha_i, \psi)|^2$, $i = 1, 2, \ldots$ of this jump) is certainly not of the type described by the time dependent Schrödinger equation. This latter always results in a continuous change of ψ, in which the final result is uniquely determined and is dependent on ψ.

Now, the projection postulate introduces a consistency problem. A measurement involves an interaction between a system, S, and a measuring instrument, M. The equation of motion of the theory describes the time-evolution of the composite system, $S + M$. Suppose we measure $S + M$ by a second measuring instrument. Then we ought to get the same result for the system S whether we apply the projection postulate directly to S, or whether we apply the projection postulate to the system $S + M$ after a suitable interaction. Von Neumann shows that this is in fact the case by a rather ingenious argument requiring a detailed examination of the quantum statistics of composite systems.

Consider two systems, S_1 and S_2, with associated Hilbert spaces $\mathcal{H}_1$ and $\mathcal{H}_2$. (The subscripts here do not denote dimensionality.) The composite system, $S_1 + S_2$, is represented in the Hilbert space $\mathcal{H}_1 \otimes \mathcal{H}_2$. Let $\{\psi_m\}$ be a complete orthonormal set of basis vectors in $\mathcal{H}_1$, and $\{\varphi_n\}$ a complete orthonormal basis in $\mathcal{H}_2$. Then $\{\Phi_{mn} = \psi_m \otimes \varphi_n\}$ is a complete orthonormal basis in $\mathcal{H}_1 \otimes \mathcal{H}_2$. S_1-magnitudes, S_2-magnitudes, and

$(S_1 + S_2)$-magnitudes are represented by self-adjoint operators in $\mathcal{H}_1$, $\mathcal{H}_2$, and $\mathcal{H}_1 \otimes \mathcal{H}_2$, respectively.

It is convenient to represent these operators by matrices. The matrix of the $\mathcal{H}_1$-operator A^1 is the array of (complex) numbers:

$$A^1_{mm'} = (\psi_m, A^1 \psi_{m'}) \quad (m, m' = 1, 2, \ldots).$$

The matrix of the $\mathcal{H}_2$-operator A^2 is the array:

$$A^2_{nn'} = (\varphi_n, A^2 \varphi_{n'}) \quad (n, n' = 1, 2, \ldots).$$

The matrix of the $\mathcal{H}_1 \otimes \mathcal{H}_2$ operator A is the array:

$$A_{mn, m'n'} = (\Phi_{mn}, A\Phi_{m'n'}) \quad (m, n, m', n' = 1, 2, \ldots).$$

Now, an S_1-magnitude may be regarded as an $(S_1 + S_2)$-magnitude. The $\mathcal{H}_1$-operator represented by the matrix

$$A^1_{mm'}$$

defines an $\mathcal{H}_1 \otimes \mathcal{H}_2$-operator represented by the matrix

$$A_{mn, m'n'} = A^1_{mm'} I^2_{nn'}$$

where $I^2_{nn'}$ is the matrix of the unit operator I in $\mathcal{H}_2$, i.e. $I_{nn'} = 1$ if $n = n'$, and $I_{nn'} = 0$ if $n \neq n'$. Similarly, the $\mathcal{H}_2$-operator represented by the matrix

$$A^2_{nn'}$$

defines an $\mathcal{H}_1 \otimes \mathcal{H}_2$-operator represented by the matrix

$$A_{mn, m'n'} = A^2_{nn'} I^1_{mm'}.$$

A statistical operator in $\mathcal{H}_1 \otimes \mathcal{H}_2$ determines a statistical operator in $\mathcal{H}_1$ and a statistical operator in $\mathcal{H}_2$, i.e. the statistical operator of the composite system $S_1 + S_2$ determines the statistical states of the sub-systems S_1 and S_2. Suppose the statistical state of the composite system is represented by the statistical operator W in $\mathcal{H}_1 \otimes \mathcal{H}_2$, with the matrix

$$W_{mn, m'n'}.$$

It is easy to show that the statistical operator W^1 defining the statistical state of the subsystem S_1 in the factor space $\mathcal{H}_1$, is represented by the matrix:

$$W^1_{mm'} = \sum_{n=1}^{\infty} W_{mn, m'n}.$$

Similarly, the statistical operator W^2 is represented by the matrix:

$$W^2_{nn'} = \sum_{m=1}^{\infty} W_{mn, mn'} .$$

Von Neumann considers the following general problem: For any two statistical matrices $W^1_{mm'}$, $W^2_{nn'}$, find a statistical matrix $W_{mn, m'n'}$ such that

$$\sum_{n=1}^{\infty} W_{mn, m'n} = W^1_{mm'}$$

$$\sum_{m=1}^{\infty} W_{mn, mn'} = W^2_{nn'} .$$

He shows that this problem has a unique solution if and only if at least one of the two matrices represents a pure statistical state. In this case the solution is:

$$W_{mn, m'n'} = W^1_{mm'} W^2_{nn'} .$$

He also shows that for any vector Φ in $\mathscr{H}_1 \otimes \mathscr{H}_2$, it is possible to choose complete orthonormal bases $\{\psi_m\}$, $\{\varphi_n\}$ in $\mathscr{H}_1$ and $\mathscr{H}_2$ such that

$$\Phi = \sum_{i=1}^{M} c_i \psi_{r_i} \otimes \varphi_{s_i}$$

with M finite or infinite, i.e. the values of the magnitudes A^1 and A^2 with eigenvectors $\{\varphi_m\}$ and $\{\psi_n\}$ are correlated in the state represented by the vector Φ:

$$p_{\Phi}(a^1 = a^1_{r_i} \ \& \ a^2 = a^2_{s_i}) = |c_i|^2$$

$$p_{\Phi}(a^1 = a^1_{r_i} \ \& \ a^2 = a^2_{s_j}) = 0 \quad (i \neq j).$$

Now, the matrix of the statistical operator $W = P_{\Phi}$ is:

$$W_{mn, m'n'} = f_{mn} f^*_{m'n'}$$

and so

$$W^1_{mm'} = \sum_{n=1}^{\infty} f_{mn} f^*_{m'n}$$

where

$$f_{mn} f^*_{m'n} = |c_i|^2 \quad \text{for} \quad m = m' = r_i; \quad i = 1, 2, \dots$$

and

$$f_{mn} f^*_{m'n} = 0 \quad \text{otherwise,}$$

i.e.

$$W^1 = \sum_{i=1}^{M} |c_i|^2 \, P_{\psi_{r_i}}.$$

Similarly:

$$W_{nn'}^2 = \sum_{m=1}^{\infty} f_{mn} f_{mn'}^*$$

where

$$f_{mn} f_{mn'}^* = |c_i|^2 \quad \text{for} \quad n = n' = s_i; \quad i = 1, 2, \ldots ,$$

and

$$f_{mn} f_{mn'}^* = 0 \quad \text{otherwise,}$$

i.e.

$$W^2 = \sum_{i=1}^{M} |c_i|^2 \, P_{\varphi_{s_i}}.$$

Thus, a *pure* statistical state $W = P_\Phi$ in $\mathcal{H}_1 \otimes \mathcal{H}_2$, projected into the factor space $\mathcal{H}_1$ or $\mathcal{H}_2$, is in general a *mixture*. (It is a pure state if $M = 1$.)

To sum up: If the systems S_1 and S_2 are represented by the vectors $\psi \in \mathcal{H}_1$ and $\varphi \in \mathcal{H}_2$, respectively, then the system $S_1 + S_2$ is represented by the vector $\Psi = \psi \otimes \varphi$ in $\mathcal{H}_1 \otimes \mathcal{H}_2$. If $S_1 + S_2$ is represented by a vector Ψ in $\mathcal{H}_1 \otimes \mathcal{H}_2$ which is not a tensor product $\psi \otimes \varphi$, then S_1 and S_2 are associated with statistical states which are mixtures. There exists an S_1-magnitude and an S_2-magnitude such that the statistical correlations determined by Ψ establish a one-one correspondence between the values of these magnitudes.

The consistency problem is resolved in the following way: Suppose we have a system S represented by the vector ψ, i.e. the pure statistical state P_ψ. Let

$$\psi = \sum_i (\alpha_i, \psi) \, \alpha_i$$

where the α_i are eigenvectors of the self-adjoint operator representing the magnitude A. By the projection postulate, measurement of the magnitude A results in a transition

$$\psi \to \alpha_i$$

with probability $|(\alpha_i, \psi)|^2$. Now consider the measurement as an interaction between the system S and a measuring instrument M suitable for measuring A. This means that S and M interact in such a way that the

Hilbert space vector representing the composite system $S + M$ during the interaction evolves (by a unitary transformation determined by the equations of motion of the theory) to a vector of the form

$$\sum_i (\alpha_i, \psi) \, \alpha_i \otimes \varphi_i.$$

The ψ_i are eigenvectors of some M-magnitude, say R, and this representation of the statistical state of $S + M$ correlates the eigenvalues of R (representing the 'pointer-readings' of the instrument M) with the eigenvalues of the magnitude A in the system S:

$$p(a = a_i \, \& \, r = r_i) = |(\alpha_i, \psi)|^2$$

and

$$p(a = a_i \, \& \, r = r_j) = 0 \quad (i \neq j).$$

In other words, if the measuring instrument is represented initially by the vector φ, then the measurement process is an interaction governed by the equation of motion of the theory which results in the transition:

$$\psi \otimes \varphi \to \sum_i (\alpha_i, \psi) \, \alpha_i \otimes \varphi_i.$$

What von Neumann shows is that there exists measuring instruments in this sense: Given a complete orthonormal set $\{\alpha_i\}$ in $\mathcal{H}_1$ (the Hilbert space of the system S) and any vector $\psi \in \mathcal{H}_1$ specifying the initial statistical state of S, there exists a complete orthonormal set $\{\varphi_j\}$ in $\mathcal{H}_2$ (the Hilbert space of M) and a vector $\varphi \in \mathcal{H}_2$ (specifying the initial statistical state of the instrument M), such that

$$\Psi(t) = \sum_i (\alpha_i, \psi) \, \alpha_i \otimes \varphi_i$$

(the statistical state vector of the composite system $S + M$ in the Hilbert space $\mathcal{H}_1 \otimes \mathcal{H}_2$) is a solution of Schrödinger's equation of motion for the composite system $S + M$, if

$$\Psi(t_0) = \psi \otimes \varphi$$

is the initial statistical state vector of the composite system. Equivalently: There exists a unitary transformation:

$$\psi \otimes \varphi \to \sum_i (\alpha_i, \psi) \, \alpha_i \otimes \varphi_i.$$

Now, by the projection postulate, a measurement of the magnitudes A and R on the composite system $S + M$ by a second measuring instrument results in the transition:

$$\Psi(t) \rightarrow \alpha_i \otimes \varphi_i$$

with probability $|(\alpha_i, \psi)|^2$, from which it follows that the statistical state vector of the system S after the interaction is α_i with probability $|(\alpha_i, \psi)|^2$. By the theory of statistical operators in $\mathcal{H}_1 \otimes \mathcal{H}_2$ developed above, if

$$W = P_\Psi, \quad \text{with} \quad \Psi = \alpha_i \otimes \varphi_i$$

then

$$W^1 = P_{\alpha_i}$$

and

$$W^2 = P_{\varphi_i}.$$

We have consistency in the following sense: The application of the projection postulate directly to the system S is consistent with its application to the system $S + M$ after a suitable interaction between S and M governed by the equation of motion of the theory.

The significance of von Neumann's projection postulate, and the resulting consistency problem and its solution, is generally misunderstood. It is not sufficient, the argument goes, merely to show consistency. What has to be shown is, firstly, that the statistical state vector of the composite system $S + M$ immediately after the measuring instrument M has registered a result corresponding to the value a_i for A is:

$$\alpha_i \otimes \varphi_i$$

and, secondly, that in a measurement of the magnitude A on the system S the transition

$$\psi \otimes \varphi \rightarrow \alpha_i \otimes \varphi_i$$

occurs with probability $|(\alpha_i, \psi)|^2$.

Now, this objection depends on the assumption that a system has an atomic property if and only if it is associated with a Hilbert space vector in the 1-dimensional subspace which is in the range of the projection operator representing the property, i.e. if and only if this projection operator is the statistical operator of the system. Thus, if the composite system $S + M$ is associated with the vector $\Psi(t)$, it cannot have the property corresponding to the value r_i of R and a_i of A, for any i: a system only has those properties assigned probability 1 by the statistical state. This assumption requires that the properties of a system incompatible with the atomic property assigned probability 1 (i.e. those properties assigned probabilities between 0 and 1) are in some sense brought into existence by the measurement process, and this is surely part of what is involved in von Neumann's proposal of the projection postulate as characterizing the peculiar disturbances of micro-systems that supposedly occur during measurement – disturbances that underly the statistical relations of quantum mechanics, according to the Copenhagen interpretation. The objection makes no sense at all if what is demanded is a reduction of the stochastic measurement transitions determined by the projection postulate to the temporal evolutions governed by the equation of motion of the theory. For, clearly, there can be no possible interaction between S and M governed by the quantum mechanical equation of motion which results in different transitions of the Hilbert space vector Ψ on different occasions. This follows immediately and trivially from the equation of motion, which is deterministic and describes a unitary transformation of the Hilbert space vector representing the composite system.

The measurement problem is usually posed as follows: that the statistical operator associated with $S + M$ after the interaction is actually

$$W' = \sum_i |(\alpha_i, \psi)|^2 \, P_{\alpha_i \otimes \varphi_i}$$

and not

$$W(t) = P_{\Psi(t)}$$

and what is to be explained is the projection postulate in the form

$$W \to W'$$

where W is the statistical operator of the composite system before the interaction. This is a weaker requirement, because the decomposition of W' as a sum of projection operators onto the 1-dimensional subspaces defined by the vectors $\alpha_i \otimes \varphi_i$ is unique only if the probabilities $|(\alpha_i, \psi)|^2$ are all distinct.

It is not at all evident just what problem would be solved by a theoretical explanation of the transition $W \to W'$ in a measurement process, for this transition could not be associated in an unambiguous way with a measurement of the magnitude A (on the above assumption that the system has an A-value if and only if the Hilbert space vector is an eigenvector of A) unless the probabilities $|(\alpha_i, \psi)|^2$ – the eigenvalues of the operator W' – were all distinct, and this would depend on ψ. But even supposing that this difficulty could be avoided in some way, the only conceivable explanation for a statistical operator representing a mixture resulting from an interaction (defined by a unitary transformation) would be to suppose that the initial statistical operator associated with the measuring instrument represented a mixture.

Now this explanation is obviously untenable, as von Neumann has shown. For suppose the statistical operator associated with M before the interaction is:

$$W^2 = \sum_i w_i P_{\varphi_i}.$$

As before, we assume that the statistical operator associated with S before the interaction is:

$$W^1 = P_\psi.$$

Then the initial statistical operator of the composite system is:

$$W = \sum_i w_i P_{\psi \otimes \varphi_i}$$

A (unitary) interaction could transform W to:

$$W' = \sum_i w_i P_{\psi_i \otimes \varphi_i}.$$

But this is of no use at all, for the probabilities, w_i, depend on M and not at all on S! We require, of course, $w_i = |(\alpha_i, \psi)|^2$, and here the w_i represent, loosely, a measure of our ignorance concerning the value of the magnitude R of the measuring instrument M before the interaction.

In general, a measurement process is an interaction between two systems, S_1 and S_2, resulting in correlations between (some of) the magnitudes of S_1 and S_2, so that the value of an S_1-magnitude, say, can be inferred from the value of an appropriate S_2-magnitude, on the basis of the theory of the interaction. From the standpoint of the theoretical analysis of measurement, the value of the S_2-magnitude is simply stipulated, i.e. the value of the S_2-magnitude is ascertained by looking at S_2, and this must be understood as 'mere looking'. Whether or not the systems are macro-systems or microsystems is irrelevant here, for it would be absurd to suppose that S_2 is disturbed in any way at all by the assignment of values to its magnitudes in this sense: the value of the S_2-magnitude is not brought into being by 'looking', or altered by 'looking'. Of course, the process of 'looking' can be treated as an actual measurement process, i.e. an interaction similar to the S_1–S_2 interaction establishing correlations between the magnitudes of S_2 and the magnitudes of a system S_3, but this introduces nothing new. For in the theoretical analysis of the measurement process the values of the S_3-magnitudes are now stipulated, i.e. they are ascertained by 'mere looking'.

For example, consider the composite classical mechanical system, $S_1 + S_2$, whose states are represented by the points $(q_1, p_1; q_2, p_2)$ in phase space, a 4-dimensional space constructed as the Cartesian product of the phase spaces of S_1 and S_2. The Hamiltonian function

$$H = q_1 p_2 + q_2 p_1$$

correlates the state of S_1 (represented by the values of the position and momentum variables q_1, p_1) to the state of S_2 (represented by the values of the position and momentum variables q_2, p_2). For Hamilton's equations of motion for the systems are:

$$\frac{dq_1}{dt} = \frac{\partial H}{\partial p_1} = q_2$$

$$\frac{dq_2}{dt} = \frac{\partial H}{\partial q_1} = q_1$$

$$\frac{dp_1}{dt} = -\frac{\partial H}{\partial q_1} = -p_2$$

$$\frac{dp_2}{dt} = -\frac{\partial H}{\partial q_2} = -p_1$$

with the solutions

$$q_1 = Ae^t + Be^{-t} \qquad q_2 = Ae^t - Be^{-t}$$
$$p_1 = Ce^t + De^{-t} \qquad p_2 = -Ce^t + De^{-t}$$

where A, B, C, D are four arbitrary constants of integration. Evidently, the state of S_1 may be inferred by looking at the state of S_2. Or, two observations (in the sense of 'mere looking') of the magnitude q_2 (or p_2) at two different times suffice to determine the magnitude q_1 (or p_1) at any time.

A similar example in quantum mechanics is the composite system consisting of two spin-$\frac{1}{2}$ particles in the singlet spin state, the system considered in Chapter VI. By looking at one particle, i.e. by ascertaining whether or not the system has a particular atomic property, we can infer whether or not the other particle has the correlated property. The fact that the properties of microsystems cannot be directly observed is quite irrelevant here. Von Neumann's measurement interaction between S and M results in a correlation between the values of the magnitudes A and R expressed by the Hilbert space vector:

$$\Psi(t) = \sum_i (\alpha_i, \psi)\, \alpha_i \otimes \varphi_i.$$

By ascertaining that the value of the M-magnitude R is r_i, say, we infer that the value of the S-magnitude A is a_i. There is no measurement problem here peculiar to quantum mechanics.

To argue that the Hilbert space vector of the composite system ought to be

$$\alpha_i \otimes \varphi_i$$

if the value of the M-magnitude R is r_i and the value of the S-magnitude A is a_i, is to assume that a quantum mechanical magnitude has a value for a system if and only if the statistical state of the system is represented by an eigenvector of the magnitude. And this assumption could only be incorporated into the Hilbert space theory of quantum mechanics by a special measurement postulate of the sort proposed by von Neumann. If the statistical state of a system is represented by the Hilbert space vector

$$\psi = \sum_i (\alpha_i, \psi)\, \alpha_i$$

then measurement of the magnitude A yielding the value a_i must involve a transition of the Hilbert space vector:

$$\psi \to \alpha_i.$$

This is von Neumann's projection postulate. Given this assumption on the quantum mechanical magnitudes (which I have associated with the Copenhagen interpretation), the only remaining problem is the consistency of the measurement postulate with the equation of motion of the theory. In order to show consistency, we assume an interaction between the system S and a measuring instrument M which establishes correlations between an S-magnitude and an M-magnitude, of the sort given by the Hilbert space vector

$$\Psi(t) = \sum_i (\alpha_i, \psi) \, \alpha_i \otimes \varphi_i$$

i.e. by the statistical operator

$$W(t) = P_{\Psi(t)}.$$

The objection that the statistical operator of the composite system $S + M$ after the interaction is actually

$$W' = \sum_i |(\alpha_i, \psi)|^2 \, P_{\alpha_i \otimes \varphi_i}$$

confuses the consistency problem with the measurement problem which is already resolved by the incorporation of the measurement postulate into the theory.

To sum up: To a certain extent, von Neumann took the disturbance theory of measurement of the Copenhagen interpretation seriously, and thought it necessary to 'complete' the Hilbert space theory of quantum mechanics by a postulate describing the peculiar stochastic transition of a microsystem under a measurement acting as a disturbance of a certain kind. This introduces a problem of consistency with the equation of motion of the theory, which von Neumann solved. The standard 'measurement problem' rests on a misunderstanding of von Neumann's problem, and a failure to see the significance of the projection postulate as the theoretical principle characterizing measurement disturbances, which according to the Copenhagen interpretation underly the statistical relations of quantum mechanics. This problem is a pseudo-problem.

But von Neumann's problem is also a pseudo-problem, since it derives from an uncritical acceptance of certain tenets of the Copenhagen interpretation, and a consequent inadequate analysis of the completeness problem. What is to be explained is the significance of statistical states which assign unit probability to atomic propositions and non-zero probabilities to other incompatible atomic propositions. Thus, it is the significance of the compatibility relation which is at issue. The probabilities

$$|(\alpha_i, \psi)|^2$$

associated with the values a_i of a magnitude A by the Hilbert space vector ψ reflect the incompatibility of the Boolean subalgebra $\mathcal{B}_A$ with any Boolean subalgebra containing a proposition represented by the 1-dimensional subspace $\mathcal{K}_\psi$. The system always has an A-property, even when the statistical state is P_ψ, and this A-property is ascertained by 'looking'. The point is that having an A-property, B-property, etc., does not require the existence of a 2-valued homomorphism on the propositional structure; the existence of 2-valued homomorphisms is associated only with Boolean propositional structures.

CHAPTER XII

THE INTERPRETATION OF QUANTUM MECHANICS

What constitutes an *interpretation* of a theory like quantum mechanics? What problem is solved by proposing an interpretation?

Einstein has introduced a distinction between *principle* theories and *constructive* theories, which is of fundamental importance for the theory of theories. In the case of constructive theories, the idea is to reduce a wide class of diverse systems to component systems of a particular kind. The existence claims to which theories of this type have led are well known. The molecular hypothesis of the kinetic theory of thermodynamic systems is an example. Classical discussions of the reality of theoretical concepts have focussed on constructive theories (Einstein (b), p. 54):

When we say that we have succeeded in understanding a group of natural processes, we invariably mean that a constructive theory has been found which covers the processes in question.

Principle theories have a different aim. These theories introduce abstract structural constraints which events are held to satisfy. Einstein's example is a theory like classical thermodynamics, which specifies (Einstein (b), p. 54)

general characteristics of natural processes, principles that give rise to mathematically formulated criteria which the separate processes or the theoretical representations of them have to satisfy.

The special and general theories of relativity are principle theories – this conception of the significance of the relativity principle is crucial for a proper understanding of the theories. As a principle theory of space-time structure, Newtonian mechanics in the absence of gravitation represents the 4-dimensional geometry of space-time by the inhomogeneous Galilean group, which acts transitively in the class of free motions, i.e. the inhomogeneous Galilean group is the symmetry group of the free motions: it is a subgroup of the symmetry group of every mechanical system, and the largest such subgroup. Einstein's special principle of relativity is the hypothesis that the symmetry group of the free motions is the Poincaré group. The transition from the Galilean group to the Poincaré group is associated with a corresponding modification in space-time structure. The absolute time and Euclidean metric of Newtonian mechanics are dropped, and the metrical relations of space-time are determined by the Minkowski tensor.

By an interpretation of a theory, I mean an account that shows in what respects the theory is related to preceding theories. In the case of principle theories, this requires a characterization of the theory as involving a modification of certain specific structural principles, representing the transition from one class of possible structures to another. I understand general relativity as a principle theory of space-time structure involving the hypothesis that the symmetry group of the free motions is the group of all diffeomorphisms, i.e. the group preserving only the local differential and topological structure of the space-time manifold. The space-time metric plays a dynamical role and is no longer an absolute element in the description of motion. An opposing interpretation is implicit in the Wheeler-Misner theory of geometrodynamics. Space is the dynamical element of geometrodynamics. On this view, general relativity is a constructive theory of matter: material systems are constructed out of the temporal behaviour of space.

The central foundational problem of quantum mechanics is the completeness problem, the problem of hidden variables. Consider, again, a system S associated with a 2-dimensional Hilbert space $\mathcal{H}_2$ (say a spin-$\frac{1}{2}$ particle). Let A and B be two incompatible magnitudes (spins) with eigenvalues a_1, a_2 and b_1, b_2, respectively. The corresponding eigenvectors are α_1, α_2 and β_1, β_2. As usual, I denote the atoms (atomic propositions or events) in the maximal Boolean subalgebras $\mathcal{B}_A$ and $\mathcal{B}_B$ of $\mathcal{A}_2$, by a_i and b_j. The statistical state associated with the vector α_1 assigns probabilities

$$p_{\alpha_1}(a_1) = 1, \qquad p_{\alpha_1}(a_2) = 0$$

to the atomic propositions in $\mathcal{B}_A$, and probabilities

$$p_{\alpha_1}(b_1) = |(\beta_1, \alpha_1)|^2, \qquad p_{\alpha_1}(b_2) = |(\beta_2, \alpha_1)|^2$$

to the atomic propositions in $\mathcal{B}_B$. The problem concerns the significance of the measures $p_{\alpha_1}(b_1)$, $p_{\alpha_1}(b_2)$, which are neither 0 nor 1. The measure $p_{\alpha_1}(b_j)$ cannot be interpreted as the conditional probability, $p(b_j | a_1)$, that the proposition b_j is true (or the corresponding event obtains) given that the proposition a_1 is true, i.e. as the probability that the value of the magnitude B is b_j given that the value of the magnitude A is a_1. For a_1 and b_j are atoms in $\mathscr{A}_2$, and so $p_{\alpha_1}(b_j)$ cannot be understood as the relative measure of the set of ultrafilters containing b_j in the set of ultrafilters containing a_1.

To put this another way: The B-statistics defined by α_1 cannot be represented by a mixed ensemble consisting of a fraction $|(\beta_1, \alpha_1)|^2$ of β_1-systems and a fraction $|(\beta_2, \alpha_1)|^2$ of β_2-systems, for this ensemble does not generate the A-statistics specified by α_1. The statistical operator

$$W = |(\beta_1, \alpha_1)|^2 P_{\beta_1} + |(\beta_2, \alpha_1)|^2 P_{\beta_2}$$

generates the same statistics as the operator P_{α_1} for all magnitudes compatible with B, but not for magnitudes compatible with A.

Now, if we assume that the properties of a system are necessarily structured in a Boolean algebra, the natural move is to regard the atomic character of a_i and b_j as spurious. In other words, because only a Boolean structure makes sense for the properties of a system, quantum mechanics is incomplete, and the statistical relations of the theory reflect an averaging process over variables whose precise values remain 'hidden' at present. This interpretation is possible for $\mathscr{A}_2$ since 2-valued homomorphisms do exist on $\mathscr{A}_2$: a_i and b_j can be represented as non-atomic properties in a Boolean algebra, i.e. a Boolean imbedding of $\mathscr{A}_2$ is possible. Thus, the B-statistics defined by α_1 can be represented by a mixture which also generates the A-statistics, if the quantum mechanical description is extended by the introduction of additional parameters – hidden variables – which enable the construction of a measure space whose 1-point subsets represent atoms in a Boolean algebra.

The special feature of a 2-dimensional Hilbert space which allows this possibility is that all magnitudes are non-degenerate (or 'maximal'). Alternatively, compatibility reduces to orthogonality – two different propositions are compatible if and only if they are represented by orthogonal subspaces in $\mathscr{H}_2$, i.e. if and only if they negate each other. In higher-dimensional Hilbert spaces, the existence of degenerate magnitudes (or non-trivial instances of compatibility) prevents a hidden variable extension of the theory which preserves the algebraic structure of the magnitudes. In the case of a degenerate magnitude B which is compatible with each of two mutually incompatible magnitudes, A and C, it is impossible to represent A, B, and C by three random variables on a measure space in such a way that the compatibility relations are preserved. We cannot introduce a functional relationship between f_A and f_B (via a random variable f_D) and a functional relationship between f_B and f_C (via a random variable f_E), without also introducing a relationship between f_A and f_C. (See Chapter VII.) *But the incompatibility of A and C requires the complete independence of A and C.*

The simplest way to see this is to notice that in the 2-dimensional case we can, formally, introduce a different measure space for each magnitude A, and define a probability measure ρ_W on X_A for each statitical state W of quantum mechanics, generating the A-statistics specified by W as the measure of points in X_A which a random variable f_A maps onto the value a_i of A:

$$p_W(a = a_i) \ = \ \rho_W(f_A^{-1}(a_i)).$$

The spaces X_A can then be combined into a Cartesian product measure space

$$X = \prod_{A \in \mathscr{Q}} X_A$$

with an appropriate product measure which generates the statistics defined by W for any magnitude. In the case of higher-dimensional Hilbert spaces, if we introduce in a similar way a different measure space for each non-degenerate magnitude, then the representation of a degenerate magnitude as a random variable is *non-unique*: each degenerate magnitude will have to be represented by a family of random variables, a different real-valued function on each measure space associated with a non-degenerate magnitude compatible with the degenerate magnitude. We cannot introduce a special measure space for each degenerate magnitude, for this would amount to treating a degenerate magnitude B, compatible with the non-degenerate magnitude A, as completely independent of A.

Thus, the construction of a classical probability space for a quantum mechanical system as the product of factor spaces associated with the maximal Boolean subalgebras in $\mathscr{A}$ is possible for $\mathscr{A}_2$ and impossible for $\mathscr{A}_3$, $\mathscr{A}_4$, ... (assuming the propositional structure is preserved in the construction), because *all* the Boolean subalgebras of $\mathscr{A}_2$ are maximal (each consisting of 0, 1, and two orthogonal atoms), while there exist non-maximal Boolean subalgebras in $\mathscr{A}_3$, $\mathscr{A}_4$,.... Each such non-maximal Boolean subalgebra, associated with a degenerate magnitude, can be extended to a maximal Boolean subalgebra in an infinite number of ways, corresponding to the family of maximal Boolean subalgebras associated with the non-degenerate magnitudes compatible with the degenerate magnitude in question. The non-uniqueness of these extensions makes a Boolean imbedding of the partial Boolean algebra impossible.

Now, it is precisely the existence of degenerate magnitudes, i.e. non-trivial instances of compatibility, which characterizes the non-Boolean propositional structure of quantum mechanics, and the peculiar statistical relations of the theory. Hidden variable theories are Boolean reconstructions of the quantum statistics which alter the propositional structure by replacing degenerate magnitudes by appropriate sets of magnitudes. Thus, magnitudes which are equivalent in the partial Boolean algebra of magnitudes of a quantum mechanical system are treated as inequivalent in a hidden variable theory: they are statistically equivalent only for those probability measures corresponding to the statistical states of quantum mechanics. Insofar as statistical ensembles are possible which do not generate the quantum statistics, such theories differ in content from quantum mechanics, and the difference is in principle testable.

For example, in the theory of Bohm and Bub a system is characterized by an ordered pair of vectors: a vector $\psi \in \mathscr{H}$ representing the (pure) statistical state of the system, and a vector $\xi \in \mathscr{H}'$ representing the additional 'hidden variables'. An equation of motion is proposed for ψ which involves ξ, and differs from Schrödinger's equation by a term which is assumed to dominate when the macroscopic environment is such as to constitute a measuring instrument for a *non-degenerate* magnitude. If the magnitude in question is A, with eigenvectors α_1, α_2, ..., α_n, then this additional term describes a process in which ψ is projected onto a particular eigenvector α_i. It follows from the equation that the resulting eigenvector is determined by the greatest ratio $|(\alpha_i, \psi)|^2 / |(\alpha_i, \xi)|^2$. Thus, the Bohm-Bub hidden variable theory involves a measure space $X = \mathscr{H} \times \mathscr{H}'$, random variables f_A on X defined for each non-degenerate magnitude by the algorithm

$$f_A(\psi, \xi) = a_i \quad \text{if} \quad \frac{|(\alpha_i, \psi)|^2}{|(\alpha_i, \xi)|^2} > \frac{|(\alpha_j, \psi)|^2}{|(\alpha_j, \xi)|^2} \quad \text{for all} \quad j \neq i,$$

and a probability measure ρ_ψ on X defined as the product of an atomic measure on $\mathcal{H}$ concentrated at the point ψ (where ψ is the statistical state) and a measure which is uniform on the surface of the unit hypersphere in $\mathcal{H}'$ and zero elsewhere on $\mathcal{H}'$. It is not difficult to show that the measure of the set of points satisfying the above inequality is in fact $|(\alpha_i, \psi)|^2$. (See Bohm and Bub for a simple proof in the 2-dimensional case; Bub for a concise generalization to the n-dimensional case. A similar theory was first proposed by Wiener and Siegel.) The problem that arises is the definition of the random variables for degenerate magnitudes, and evidently there is no way of defining these variables without introducing inequivalent representation for magnitudes which are equivalent in the partial algebra.

A hidden variable theory will yield probabilities which differ from those generated by the quantum algorithm for measures which do not correspond to the statistical states of quantum mechanics. Statistical ensembles represented by such measures are always in principle possible in a hidden variable theory, and generally will have to be treated as 'unstable' by introducing some mechanism which tends to transform these exceptional statistical ensembles to 'equilibrium' ensembles represented by measures corresponding to quantum statistical states. For example, the Bohm-Bub theory postulates a randomization process which tends to reconstitute the random measure over the unit hypersphere in the Hilbert space $\mathcal{H}'$, the only measure yielding the quantum statistics.

Now, a theory constructed along these lines might very well be interesting for all sorts of reasons. (It might, for example, be true.) But the replacement of quantum mechanics by an alternative theory cannot be regarded as a contribution to the foundational problem of interpretation, for the problem is simply ignored. Hidden variable theories in this sense will have to stand on their own feet.

Heisenberg's version of the Copenhagen interpretation (as I have reconstructed it in Chapter II) is the interpretation of quantum mechanics as a 'degenerate' hidden variable theory. The statistical relations of quantum mechanics are understood with reference to a Boolean reconstruction in which the only statistical ensembles that can be constructed physically (by measuring instruments in principal at our disposal) are those represented by measures corresponding to the statistical states of quantum mechanics. In terms of the Bohm-Bub theory: the randomization process is always infinitely fast, i.e. the randomization time is zero.

Bohr's dispositional interpretation of the quantum mechanical magnitudes involves the assumption that a (micro-) system, at any one time, is characterized by a set of properties which form a Boolean algebra, corresponding to a maximal Boolean subalgebra in the partial Boolean algebra of quantum mechanical idempotent magnitudes. The appropriate Boolean subalgebra is related to the macroscopic measuring instruments, which define conditions appropriate for the realization of some set of dispositions for the system to manifest a family of properties forming a Boolean algebra. In effect, this amounts to saying that a system is always represented by a mixture of quantum states (in the limiting case by a mixture with 0, 1 weights), the constituents of the mixture being determined by the experimental conditions. If the experimental conditions are such as to determine the maximal Boolean subalgebra $\mathscr{B}_A$ associated with the non-degenerate magnitude A, then the system is actually in one of the states α_i, and hence represented by a statistical operator of the form

$$\sum_i w_i P_{\alpha_i}.$$

If the state is known, $w_i = 1$ or 0. The probabilities assigned to the 'complementary' properties in $\mathscr{B}_B$ by a pure statistical operator P_{α_i}, representing a statistical ensemble of systems all in the state α_i, are to be understood as the probabilities of finding particular B-values, if the experimental conditions are altered so as to determine the maximal Boolean subalgebra $\mathscr{B}_B$, given that the system is in the state α_i. The quantum mechanical description of a system, in terms of a partial Boolean algebra of dispositions, allows the consideration of all possible experimental conditions, but the application of this description to a particular system is always with respect to the experimental conditions obtaining for the system, which specify a particular maximal Boolean subalgebra of properties, via the dispositions for the system to manifest such properties under these conditions.

Now this interpretation leads to an insoluble measurement problem, if the experimental conditions for a system S are assumed to be determined (in principle, at least) by a physical interaction between S and a second system M (which, even if macroscopic, ought to be reducible to a complex of interacting microsystems). There is no way in which a particular maximal Boolean subalgebra $\mathscr{B}_A$ can be selected for S by an interaction with M governed by the quantum mechanical equation of motion, for mechanical interactions are unitary transformations which do not violate the integrity of the Hilbert space: a vector can always be represented with respect to any orthogonal set of basis vectors. The only possible way in which quantum mechanics can be made to fit this interpretation – as von Neumann saw – is by including a special projection postulate in the theory, which has the effect of 'filtering out' a suitable maximal Boolean subalgebra in each case of measurement.

From this point of view, there is nothing to choose between Bohr's version of the Copenhagen interpretation and Heisenberg's, for the introduction of the projection postulate is quite analogous to the hypothesis of a discontinuous, infinitely fast randomization process for the hidden variables. This is like interpreting classical electrodynamics as an ether theory supplemented by Lorentz's length contraction and time dilatation postulate, as opposed to Einstein's interpretation in the special theory of relativity.

Fundamentally, then, there is only one puzzle about the quantum statistics, which arises because of the non-existence of 2-valued homomorphisms on the partial Boolean algebra of propositions, i.e. the breakdown of the one-one correspondence between atomic events, ultrafilters, and 2-valued homomorphisms which holds for Boolean algebras. The variety of conceptual puzzles and paradoxes (the 2-slit experiment, tunnelling paradox, Schrödinger's cat, etc. – 'interference' phenomena of one sort or another) all reflect the impossibility of interpreting the non-zero probabilities between incompatible atoms as conditional probabilities. A paradox is generated by requiring a non-Boolean propositional structure to satisfy certain conditions which only a Boolean structure can satisfy (for example, that all ultrafilters are prime filters, or that a proposition can only be true or false relative to a truth value assignment to all the propositions defined by a 2-valued homomorphism).

In the previous chapters, I have represented classical and quantum mechanics as principle theories of logical structure, since they introduce constraints on the way in which the properties of a physical system are structured. The logical structure of a physical system is understood as imposing the most general kind of constraint on the occurrence and non-occurrence of events. I have argued that the transition from classical to quantum mechanics is to be understood as a generalization of the

Boolean propositional structures of classical mechanics to a particular class of non-Boolean structures. There are two aspects to this thesis: firstly, the significance of a realist interpretation of logical structure analogous to Einstein's realist interpretation of geometric structure involved in the transition from classical to relativistic physics; secondly, the resolution of problems of interpretation by relating the peculiar statistical relations of quantum mechanics to the specific character of the underlying logical structure. Insofar as problems of interpretation remain, they are either problems about logic, or problems about the category of algebraic structures in relation to the Boolean structures of classical mechanics.

Both the search for alternative hidden variable theories of the microlevel and the Copenhagen disturbance theory of measurement (the Bohrian dispositional interpretation of the quantum mechanical magnitudes) misconstrue the foundational problem of interpretation by introducing extraneous considerations which are completely unmotivated theoretically. Thus, it may well be the case that there are features of microsystems which quantum mechanics, in its present form, cannot explain, so that the theory will have to be drastically revised by the introduction of new parameters, or perhaps even scrapped altogether. But this has no bearing on the theoretical significance of the transition from classical to quantum mechanics. The Copenhagen interpretation grounds the completeness of quantum mechanics (in the sense of the impossibility of a hidden variable theory of micro-events with a non-zero randomization time) on a thesis concerning the peculiarities of measurement at the microlevel, and this, too, is irrelevant to the problem of interpretation. For, while a thesis of this sort would explain why the description of the microlevel requires an irreducibly statistical theory of a certain kind, it cannot guarantee that quantum mechanics does in fact have this character, nor does it answer the question of how the quantum theory is related to other statistical theories. Both interpretations stem from an uncritical approach to foundational issues, and an inadequate theory of logical structure.

BIBLIOGRAPHY

Bell, J. S.: (a) 'On the Problem of Hidden Variables in Quantum Mechanics', *Reviews of Modern Physics* **38** (1966) 447–475; (b) 'On the Einstein-Podolsky-Rosen Paradox', *Physics* **1** (1964) 195–200.

Bell, J. L., and Slomsen, A. B.: *Models and Ultraproducts: An Introduction*, North Holland Publishing Company, Amsterdam, 1969.

Birkhoff, G. and von Neumann, J.: 'The Logic of Quantum Mechanics', *Annals of Mathematics* **37** (1936) 823–843.

Bohm, D.: 'On the Role of Hidden Variables in the Fundamental Structure of Physics', in *Quantum Theory and Beyond* (ed. by T. Bastin), Cambridge University Press, Cambridge, 1971.

Bohm, D. and Bub, J.: 'A Proposed Solution of the Measurement Problem in Quantum Mechanics by a Hidden Variable Theory', *Reviews of Modern Physics* **38** (1966) 453–469.

Bohr, N.: (a) 'Can Quantum-Mechanical Description of Physical Reality Be Considered Complete?', *Physical Review* **48** (1935) 696–702; (b) 'Discussion with Einstein on Epistemological Problems in Atomic Physics', in *Albert Einstein: Philosopher-Scientist* (ed. by P. A. Schilpp), Harper Torchbooks, New York, 1959, pp. 199–244.

Bub, J.: 'What is a Hidden Variable Theory of Quantum Phenomena?', *International Journal of Theoretical Physics* **2** (1969) 101–123.

Clauser, J. F., Horne, M. A., Shimony, A., and Holt, R. A.: 'Proposed Experiment to Test Local Hidden Variable Theories', *Physical Review Letters* **23** (1969) 880.

Einstein, A.: (a) Letter to Schrödinger, dated May 31, 1928 in *Letters on Wave Mechanics* (ed. by K. Przibram, transl. by M. J. Klein), Philosophical Library, New York 1967; (b) 'What is the Theory of Relativity?', in *Essays in Science*, Philosophical Library, New York. (Originally published in the London *Times*. Also reprinted in Einstein's *Ideas and Opinions*, Crown Publishers, New York, 1963.)

Einstein, A., Podolsky, B., and Rosen, N.: 'Can Quantum-Mechanical Description of Physical Reality Be Considered Complete?', *Physical Review* **47** (1935) 777–780.

Finkelstein, D.: 'The Logic of Quantum Physics', *Transactions of the New York Academy of Science* **25** (1962–63) 621–637.

Freedman, S. J. and Clauser, J. F.: 'Experimental Test of Local Hidden-Variable Theories', *Physical Review Letters*, 1972.

Gleason, A. M.: 'Measures on the Closed Subspaces of a Hilbert Space', *Journal of Mathematics and Mechanics* **6** (1957) 885–893.

Heisenberg, W.: 'Quantum Theory and Its Interpretation', in *Niels Bohr* (ed. by S. Rozental), Interscience Publishers, New York, 1967.

Holt R. A.: 'Atomic Cascade Experiments', docteral dissertation 1973, Harvard University, unpublished.

Jauch, J. M.: *Foundations of Quantum Mechanics*, Addison-Wesley Publishing Company, U.S.A., 1968.

Jauch, J. M. and Piron, C.: 'Can Hidden Variables Be Excluded in Quantum Mechanics?', *Helvetia Physica Acta* **36** (1963) 827–837.

Kochen, S. and Specker, E. P.: 'The Problem of Hidden Variables in Quantum Mechanics', *Journal of Mathematics and Mechanics* **17** (1967) 59–87.

Kolmogorov, A. N.: *Foundations of the Theory of Probability*, Chelsea Publishing Company, New York, 1950.

London, F. and Bauer, E.: *La Theorie de l'Observation en Mecanique Quantique*, Hermann et Cie, Paris, 1939.

Petersen, A.: 'The Philosophy of Niels Bohr', *Bulletin of the Atomic Scientists*, September, 1963, pp. 8–14.

Sikorski, R.: *Boolean Algebras*, Springer-Verlag, Berlin, 1964.

von Neumann, J.: *Mathematical Foundations of Quantum Mechanics*, Princeton University Press, Princeton, 1955.

Wiener, N. and Siegel, A.: 'The Differential-Space Theory of Quantum Systems', *Il Nuovo Cimento (Supplemento al Volume II, Serie X)* (1955) 982–1003.

Wigner, E. P.: 'On Hidden Variables and Quantum Mechanical Probabilities', *American Journal of Physics* **38** (1970) 1005–1009.

INDEX OF SUBJECTS

Atom, 105, 106, 120ff., 143, 144
Atomic event, 78, 94, 106, 120, 121, 149
Atomic formula, 95
Atomic proposition, 94, 121, 124, 125, 143
Axiom, 95
Axiom of Choice, 104

Basis, 17, 21, 24
Bell's inequality, 78, 79, 82ff.
Biconditional, 94, 111ff.
Bohm-Bub hidden variable theory, 146,
    147
Bohr-Kramers-Slater theory, 129
Boolean algebra, 56, 61, 93, 99ff., 144
Boolean function, 108ff.
Borel set, 3, 20ff., 34, 35
Born's probabilistic interpretation, 19
Bound variable, 95
Bounded linear operator, 14

Cartesian coordinate system, 2, 10
Characteristic function, 46, 105
Closed linear manifold, 12
Closed linear operator, 14
Commeasurability, 67
Commutation relation, 3, 36
Commutative algebra, 68
Commute, 28, 29
Compatibility (compatible), 28ff., 57, 58,
    67, 68, 84ff., 144
Complement, 56
Complementarity (complementary), 42ff.
Completeness (complete)
    Einstein-Podolski-Rosen condition of,
        38, 39
    of a Hilbert space, 11
    of a logic, 96, 100, 104
    of quantum mechanics, VIII, 38, 46ff.,
        49ff., 63, 72, 78, 85ff., 143, 150
Complete orthonormal set, 11, 12, 16, 17,
    23

Complete set of states, 88
Compton-Simon experiment, 129
Conditional, 94
Conditional probability, 77ff., 120, 121,
    143
Conjunction, 93
Consistent, 100
Constructive theory, VIII, 142
Continuous function, 11
Continuous linear operator, 14
Continuous spectrum, 21, 24
Copenhagen interpretation, VII, 37ff., 49,
    128, 140, 147
Convex set, 24, 25, 26

Definite linear operator, 25, 26, 27
Degenerate eigenvalue, 16, 17
Degenerate linear operator, 16, 17, 28
Degenerate magnitude, 2, 144, 147
Determinant, 16
Dimension, 9, 10
Dirac δ-function, 6, 8
Dirac-Jordan transformation theory, 5, 6
Discrete spectrum, 21, 23
Disjunction, 95
Dispersion, 35, 36, 47, 60
Dispersion-free, 36, 51ff., 60
Distribution function, 122
Distributive lattice, 56, 57
Disturbance theory of measurement, 37,
    42, 45, 53, 128, 140

Eigenfunction, 5
Eigenvalue, 4, 16, 18
Eigenvalue equation, 4, 23
Eigenvalue problem, 4
Eigenvector, 4, 16, 17, 18
Einstein-Podolsky-Rosen paradox, 38ff.
Everywhere dense, 11, 14
Expectation value, 18, 22, 24, 25, 34, 46,
    123

Field, 33, 103, 104
Filter, 100, 101
First-order logic, 92

Generalized probability measure, 89
Gleason's theorem, 88, 89, 90

Hamiltonian, 3, 138
Hermitian, 15
Hidden variables, VIII, 46ff., 49, 53, 65,
    88, 128, 143ff.
Hilbert space, 2, 7, 8
Homogeneous ensemble, 50ff.
Homogeneous statistical state, 26, 28
Homomorphism, 68, 101, 102, 112
Hypermaximal, 15, 21

Idempotent, 13, 123, 125
Imbeddability (imbedding), 68, 71, 87,
    112, 113, 116, 117
Incompatible, 31, 35, 145
Incompleteness (incomplete), 47, 126
Inconsistent, 99, 100
Inference rule, 94
Infimum, 55
Interpretation, VII, 141ff.
    see also Copenhagen interpretation
Interpretation of a formal theory, 94, 102
Isomorphism (isomorphic), 101, 102, 104

Join, 56

Kernel, 6

Lattice, 55ff.
Law of large numbers, 122, 125
Lebesgue-Stieltjes measure, 22
Limit point, 11, 12
Lindenbaum-Tarski algebra, 93, 97, 99,
    104, 105
Linear independence, 9, 10
Linear manifold, 10
    closed linear manifold, 12
Linear operator, 13, 14
    bounded, 14
    closed, 14
    continuous, 14
    defined everywhere, 13
    definite, 25, 26, 27

degenerate, 2, 16, 17, 27
    extension of, 14
    Hermitian, 15
    hypermaximal, 15, 23
    maximal, 15
    self-adjoint, 15
Local hidden variable theory, 83
Locality condition, 72, 74, 76
Logically true, 95

Matrix, 3, 16, 131, 132
Matrix mechanics, 2, 3, 7, 8
Maximal linear operator, 15
    non-maximal, 16
Maximal magnitude, 144
Measurable function, 34
Measurable set, 34
Measurement problem, 136, 140, 148
Meet, 56
Mixture, 3, 26, 133
Model, 95, 102, 115, 116
Modular, 59
Modus ponens, 97
Multiplicity, 16

Non-degenerate, 16, 17, 144, 146
Non-maximal, 16
Norm, 8
Notational convertion, 8, 18, 21

Operator, see linear operator
Orthocomplement, 56
Orthogonality, 9
Orthonormal, 10, 16, 17, 23

Partial algebra, 66, 67, 68
Partial Boolean algebra, 67, 68, 89, 109
Perfect field of sets, 103
Phase space, 1, 105, 106
Prime filter, 120
Principal theory, VIII, 142, 149
Probability, 32, 33, 35, 120, 123, 128, 130,
    135, 136, 140, 149
Projection operator, 3, 13, 17, 19, 20, 21,
    25, 58, 59
Projection postulate, 128, 130, 135, 136,
    140, 149
Proof theory, 95
Proper filter, 101

Proposition, 55
Propositional calculus, 96
Propositional function, 108
Pure statistical state, 2, 26, 38, 133

Q-valid, 111, 116
Quantifier, 93

Random variable, 33, 34
Reduced field of sets, 103
Refutable, 110
Relational structure, 94

Satisfaction (satisfied), 94, 95
Scalar product, 8, 20
Schmidt orthonormalization procedure, 12
Schrödinger's wave equation, 5, 19
Secular equation, 16
Self-adjoint, 13, 16, 17, 19, 23
Semi-simplicity, 113
Sentence, 94
Separability, 11
Sound, 96
Span, 10, 16, 23
Spectral measure, 3, 21, 23
Spectral representation, 17, 18, 19
Spectral theorem, 23
Spectrum, 16, 18
  continuous, 21, 24
  discrete, 21, 23
  simple, 18

State, 1, 106
Statistically equivalent, 122
Statistical matric, 132
Statistical operator, 25, 131
Statistical state, 18, 24
Stone isomorphism, 120
Stone's theorem, 100, 103, 104
Subspace, 12, 58, 59
Supremum, 55

Tautology, 96, 100, 113, 116, 117, 118
Tensor product, 39, 73, 124
Trace, 3, 24
Triangle inequality, 9

Ultrafilter, 78, 93, 101, 120
  determined by a point, 103
Uncertainty principle, 36, 71
Universally valid, 95
  see valid
Unitary operator, 19

Validity, 107, 108ff.
Variance, 34

Wave mechanics, 2, 5, 7
Weak imbeddability (weak imbedding), 112, 113
Weakly modular, 60

Zorn's lemma, 103, 104

THE UNIVERSITY OF WESTERN ONTARIO
SERIES IN PHILOSOPHY OF SCIENCE

A Series of Books on Philosophy of Science, Methodology, and Epistemology
published in connection with
the University of Western Ontario Philosophy of Science Programme

Managing Editor:

J. J. LEACH

Editorial Board:

1. J. LEACH, R. BUTTS, and G. PEARCE (eds.), *Science, Decision and Value.* Proceedings of the Fifth University of Western Ontario Philosophy Colloquium, 1969. 1973, vii + 213 pp.

2. C. A. HOOKER (ed.), *Contemporary Research in the Foundations and Philosophy of Quantum Theory.* Proceedings of a Conference held at the University of Western Ontario, London, Canada, 1973, xx + 385 pp.

3. J. BUB, *The Interpretation of Quantum Mechanics.* 1974, ix + 155 pp.